BIODETERIORATION
OF CONCRETE

BIODETERIORATION OF CONCRETE

Thomas Dyer

University of Dundee
Division of Civil Engineering
Dundee, Scotland, UK

CRC Press
Taylor & Francis Group
Boca Raton London New York

CRC Press is an imprint of the
Taylor & Francis Group, an **informa** business

A SCIENCE PUBLISHERS BOOK

CRC Press
Taylor & Francis Group
6000 Broken Sound Parkway NW, Suite 300
Boca Raton, FL 33487-2742

First issued in paperback 2021

ISBN-13: 978-0-367-78212-2 (pbk)
ISBN-13: 978-1-4987-0922-4 (hbk)

Library of Congress Cataloging-in-Publication Data

Names: Dyer, Thomas D., author.
Title: Biodeterioration of concrete / Thomas Dyer, University of Dundee
Division of Civil Engineering, Dundee, Scotland, UK.
Description: Boca Raton, FL : CRC Press, [2016] | "A science publishers book."
| Includes bibliographical references and index.
Identifiers: LCCN 2017005278| ISBN 9781498709224 (hardback : alk. paper) |
ISBN 9781498709231 (e-book)
Subjects: LCSH: Concrete--Biodegradation.
Classification: LCC TA440 .D94 2016 | DDC 620.1/3623--dc23
LC record available at https://lccn.loc.gov/2017005278

Visit the Taylor & Francis Web site at
http://www.taylorandfrancis.com

and the CRC Press Web site at
http://www.crcpress.com

To
Judith, Angus and Oscar

Preface

Awareness of the importance of ensuring durability of concrete has been a growing concern for engineers. There is a fairly good understanding of the mechanisms which cause its deterioration and the ways of limiting such damage through the use of appropriate materials and approaches to design.

Many of the deterioration mechanisms which affect concrete are the result of interaction with the non-living environment—chlorides in seawater, carbon dioxide in the atmosphere, cyclic freezing and thawing. However, living organisms can also cause damage—through both chemical and physical processes—which, under the right conditions, can be severe.

This book examines all forms of concrete biodeterioration together for the first time. It examines, from a fundamental starting point, biodeterioration mechanisms, as well as the conditions which allow living organisms (bacteria, fungi, plants and a range of marine organisms) to colonise concrete.

It also includes a detailed examination of chemical compounds produced by living organisms with respect to their interaction with the mineral constituents of cement and concrete, and the implications this has for the integrity of structures.

Approaches to avoiding biodeterioration of concrete are also covered, including selection of materials, mix proportioning, design, and the use of protective systems.

Contents

Chapter 1

Introduction

Biodeterioration of Concrete

Around 1899 the outfall sewer serving Los Angeles—which had been built only four years previously—started showing signs of severe deterioration [1]. The concrete lining of the sewers and the lime mortar used in the brickwork was undergoing significant expansion, before crumbling away. The bricks themselves appeared unaffected, save for those which spalled away as a result of the considerable pressure exerted by the expansion. This might have been attributed to aggressive substances dissolved in the sewage itself, were it not for the inconvenient fact that the corrosion was occurring above the waterline.

Early in the 20th century, the construction of brick-lined sewers was superseded by the use of concrete pipes. However, the problem of corrosion persisted. Corrosion above the waterline suggested that a gas was causing the corrosion and the most likely candidate was hydrogen sulphide (H_2S), a malodorous and potentially lethal gas produced by sulphate-reducing bacteria in sewage. This theory appeared to be supported by the fact that rates of deterioration were significantly reduced by measures taken to prevent the release of H_2S. In the Los Angeles case, modification of the sewers to keep them fully flooded—effectively turning them into a septic tank—solved the problem. Moreover, in 1929, reduced rates of deterioration were observed in concrete sewers in Orange County and El Centro County, California, after chlorination of the sewage was initiated [2].

However, H_2S—whilst being extremely hazardous—is actually not corrosive to hardened cement. The most likely explanation for the damage was that the gas was being converted into sulphuric acid. Chemical reactions capable of causing this conversion were proposed, but were not (and, in fact, could not) be demonstrated experimentally. The unsatisfactory nature of the proposed mechanism of acid formation led Guy Parker (a bacteriologist who was part of a team facing a similar problem in concrete sewers in

Melbourne, Australia) to consider the possibility that microorganisms were playing a role not only in the formation of H_2S, but in its conversion to sulphuric acid. Through a series of experiments, he was able to isolate a species of bacteria which was responsible, which he provisionally named *Thiobacillus concretivorus* [3]. In fact, the species was an already known species, *Thiobacillus thiooxidans*, but this did not depreciate Parker's astonishing discovery, clearly expressed in the original name: here was an organism that ate concrete.

The above narrative is almost certainly not the first case of *biodeterioration* of concrete, but it is one the most well-documented early examples. Biodeterioration has been defined as *'any undesirable change in the properties of a material caused by the vital activities of organisms'* [4]. It also highlights one of the important features often associated with biodeterioration—a mechanism which involves multiple organisms each playing individual and necessary roles.

Biodeterioration and Durability

One of the greatest challenges to engineers designing concrete structures and infrastructure is that of durability: a fundamental objective of design should be that a structure is able to satisfy its functions for the duration of its intended working life. The factors which must be considered by a designer in achieving this include a wide range of structural and non-structural aspects of the shape and dimensions of each element, as well as the suitability of the materials used and their relative proportions.

In the early years of reinforced concrete construction, it was believed by many engineers that iron or steel encased in concrete was by its nature invulnerable to many of the deterioration processes which challenged the integrity of other construction materials. This turned out to be somewhat optimistic: concrete is a porous material which will allow substances to permeate it. These substances can cause chemical and physical deterioration of the concrete itself, or may create chemical environments inside a concrete element which promote the corrosion of reinforcing steel. Moreover, the external environment of a concrete structure and the loads that are applied to it permanently or periodically may cause physical deterioration.

Our understanding of the mechanisms of deterioration of reinforced concrete has developed considerably over the past 150 years. It is now reasonably safe to say that it is entirely possible for a concrete structure to remain structurally serviceable for 150 years or more—if that is required—as long as appropriate design procedures are followed, and that the quality of materials and workmanship is high. Despite this, structures still regularly become unserviceable for reasons related to compromised durability, and when this occurs—or, ideally, before—maintenance is required.

Maintenance is a necessary aspect of the operation of any structure. However, where it is required to rectify structural deterioration within the intended working life, it represents a burden to the operator which was probably not budgeted for.

Many of the processes which compromise the durability of concrete structures are a consequence of the fact that the Earth is a living planet. The most obvious example of this is the corrosion of steel reinforcement. This process requires gaseous oxygen, whose presence is maintained in the atmosphere as a result of the photosynthesis undertaken by plants. Moreover, the presence of sulphate minerals in both seawater and soil—which can produce sulphate attack in concrete—is partly the result of the role played by microorganisms in the Earth's sulphur cycle.

A proportion of atmospheric carbon dioxide—which causes carbonation of concrete leading to the depassivation of reinforcement—is biogenic. It should be stressed that the presence of life on this planet maintains carbon dioxide at levels much lower than would be the case for an abiotic planet —one need only compare the CO_2 concentrations on Mars and Venus with those on Earth. Nonetheless, around 30% of our planet's atmospheric CO_2 comes from a single species, whose current dependency on the combustion of fossil fuels is currently generating growing environmental concern.

All of these processes are global in nature and the threat they pose to concrete durability—and the solutions available—have been covered by many other books. This author has chosen to concentrate on biodeterioration processes directly caused by living organisms in close proximity to concrete. The types of organism which can take part in the deterioration of concrete are surprisingly varied, including bacteria, fungi, both higher and lower plants, and a possibly surprising array of animals including molluscs and marine worms.

This may come as something of a surprise: concrete would seem, in theory, to be a highly sterile substance, containing little by way of the elements required by living things for energy or growth. Furthermore, concrete is often highly alkaline and, thus, presents a hostile chemical environment for life. Nonetheless, the ability of life to occupy far more challenging locations has frequently been noted, and the colonization of concrete surfaces is no exception.

The mechanisms of biodeterioration of concrete include both chemical processes—such as leaching of cement resulting from the formation of acidic substances—and physical ones—for instance, from pressures exerted by root growth. However, deterioration mechanisms can also involve fairly complex combinations of both. An example of this is the production of citric acid, which will form insoluble calcium citrate in contact with hydrated cement. The resulting precipitates generate pressures within concrete pores leading to fragmentation at the surface.

Deterioration rates vary considerably and will depend on the type and number of organisms involved, as well as the environmental conditions and the composition and properties of the concrete. However, in cases where long service lives are required, even slow rates of deterioration may be unacceptable. This is certainly true in the case of much of our infrastructure—such as bridges and sewers—but becomes fundamental in the case of containment applications such as nuclear waste storage and oil well decommissioning, where service lives of hundreds or thousands of years may be required.

This Book

This book aims to examine, as widely as possible, the different ways in which living organisms can compromise the durability of concrete, and to also explore means by which deterioration can be prevented or controlled.

Following on from the discussion above, we might be led to assume that the presence of life in close proximity to concrete should always lead to problems. This is not the case—in most instances living organisms will have little impact on the durability of a structure. Indeed, a number of instances are discussed in this book which their presence has a protective effect.

This raises the more general issue of whether it is appropriate to attempt to remove a particular species from around a structure simply to improve durability. This partly depends on what sort of organism is involved, but this is not the whole issue: the organism in question will make up part of a much larger ecosystem, and its removal may have significant impacts on the local environment as a consequence. Furthermore, whilst the use of biocides as a means of protection from biodeterioration and biofouling has been widely practiced, the secondary toxic effects to other species potentially significant distances away from the point of use have led to this approach being questioned. Legislation, plus a more general desire to act in an environmentally responsible manner, is leading to a move away from biocides. This book has been written with this trend in mind.

One aspect of the colonization of concrete structures by living organisms which this book attempts to avoid is the matter of aesthetics. This is largely because of the difficulties in making judgements about what constitutes an aesthetically disagreeable state. It is notable that opinions in this respect vary considerably. A plant climbing the side of a building is considered attractive to many, but is disagreeable to others. The discolouration of a facade by lichen, fungi or algae is considered by many to be ugly and requiring cleaning, whilst to others it is the building's 'patina'. Thus, this author has chosen to steer clear of adopting any position on this issue.

The book is divided into six chapters (including this introduction). Because chemical processes play such an important role in many forms of biodeterioration, it is necessary to understand the manner in which the most common compounds produced by living organisms interact with hardened cement and concrete. Chapter 2 does this by addressing the most common compounds separately. In many ways the chapter is of a format more suited to reference books. However, this approach is deemed necessary on the grounds that each interaction is unique to the compound in question and has very different effects in terms of the physical changes imparted to concrete. Moreover, since many compounds will be encountered in discussions of biodeterioration imparted by several different types of organism, this approach is likely to ultimately be useful to the reader.

The current three-domain system for classifying organisms, and the kingdoms contained within each domain is structured as follows:

Archaea	**Bacteria**	**Eukara**
Archaebacteria	Eubacteria	Protista
		Fungi
		Plantae
		Animalia

Chapter 3 of this book deals with biodeterioration resulting from the activities of both the Archaea and Bacteria domains together. Furthermore, it takes the somewhat retrograde step of referring to all organisms within these two domains as 'bacteria'—the current system of domains has existed since it was argued that Archaea and Bacteria were sufficiently different to justify separate classification [5]. Whilst it is recognised that there are many fundamental differences between these two domains, within the context of concrete biodeterioration it makes more sense to discuss them together than separately.

Chapters 4 and 5 cover fungi and plants. The format of Chapters 3 to 5 is similar. First, an overview of the life-cycle of each type of organism is presented, with particular emphasis on energy sources, growth and reproduction. Second, the metabolism of the organism is examined, particularly with regards to the manner in which chemical compounds are produced, and the benefits and challenges presented by concrete as a host environment considered. Potential deterioration mechanisms are discussed and factors influencing rates of deterioration examined. Such factors include both those related to concrete characteristics and environmental conditions. Finally, various approaches for limiting damage from biodeterioration are systematically considered and appraised.

Chapter 6 examines biodeterioration from animals, with emphasis placed on marine animals, where examples of damage are more widely reported. However, the potential for bird droppings to damage concrete surfaces is also evaluated.

References

[1] Olmstead FH and Hamlin H (1900) Converting portions of the Los Angeles outfall sewer into a septic tank. Engineering News and American Railway Journal **44:** pp. 317–318.

[2] Goudey RF (1928) Odor control by chlorination. California Sewage Works Journal **1:** 87–101.

[3] Parker CD (1945) The corrosion of concrete. 1. The isolation of a species of bacterium associated with the corrosion of concrete exposed to atmospheres containing hydrogen sulphide. Australian Journal of Experimental Biology and Medical Science **23:** 81–90.

[4] Hueck HJ (1965) The biodeterioration of materials as part of hylobiology. Material und Organismen **1:** 5–34.

[5] Woese C, Kandler O and Wheelis M (1990) Towards a natural system of organisms: proposal for the domains Archaea, Bacteria, and Eucarya. Proceedings of the National Academy of Science **87:** 4576–4579.

Chapter 2

The Chemistry of Concrete Biodeterioration

2.1 Introduction

Many of the biodeterioration processes described in subsequent chapters involve the production of chemical substances by organisms. These substances are, in some cases, products of excretion processes, whilst in other cases are produced with the purpose of modifying an organism's environment to its benefit. The influences of these substances on concrete are diverse, since reactions between cement hydration products in the concrete yield products whose characteristics vary widely.

Many of these substances are produced by more than one kingdom or domain, and so it makes sense to discuss the chemistry of these substances in the context of concrete first, and to subsequently use this chapter as a reference for subsequent discussion. Thus, this chapter provides information on the aqueous chemistry of various biogenic products in the presence of the most relevant elements present in concrete. It then examines, in generic terms, the various mechanisms that can lead to deterioration of concrete through chemical reaction.

Before introducing specific substances, a brief introduction is provided to the reader regarding some of the concepts used to characterize their behaviour. These are acidic strength, the formation of metal complexes in solution and the solubility product. The chapter employs a form of *solubility diagram* to illustrate the influence of the substances on cementitious systems. For this reason, an introduction to these diagrams is included, along with a brief description of the method used to construct such diagrams for this book.

2.1.1 Acidic strength

The strength of an acid is dependent on its ability to deprotonate—shed protons (H^+)—in solution. Thus, a strong acid, HA, will undergo the reaction

$$HA \rightleftharpoons A^- + H^+$$

to completion when it is dissolved in water.

For many acids, this reaction will reach equilibrium with only a proportion of molecules deprotonated. The extent to which deprotonation occurs for a given acid is expressed in terms of the *acid dissociation constant* (K_a). This constant is defined by the equation:

$$K_a = \frac{[H^+][A^-]}{[HA]}$$

With square brackets denoting concentration. For convenience, K_a is often expressed as its negative logarithm (pK_a). A strong acid is one with a pK_a value of less than –1.74.

The *pH* of the resulting solution is given be the equation:

$$pH = -\log[H^+]$$

pH and pK_a are intrinsically related, and the pH of a solution containing acid HA is described by the equation:

$$pH = pK_a + \log\frac{[A^-]}{[AH]}$$

Thus, if a base is progressively dissolved in the water in which the acid is dissolved, as the pH of the solution increases, the degree of deprotonation will increase: [A⁻] will increase at the expense of [AH]. When $pH = pK_a$ the concentrations of protonated and unprotonated acid molecules will be the same and the ratio [A⁻]/[AH] will equal one. As pH increases further, the solution will gradually approach a state in which all of the acid molecules are deprotonated.

The proportion of molecules of an acid in a given state of deprotonation is expressed as the *abundance ratio* (*α*). This is the concentration of the species expressed as a ratio of the total concentration of all species present. Where the species is the deprotonated acid, the abundance ratio is referred to as the *dissociation ratio*, described by the equation:

$$\alpha = \frac{[A^-]}{[A^-] + [AH]}$$

The change in abundance ratios of the deprotonated and protonated species of acetic acid is illustrated in Figure 2.1 using acetic acid (pK_a = 4.76) as an example.

Acids may possess multiple protons, each of which will have an acid dissociation constant. For example, the pK_a values for citric acid are 3.13, 4.76 and 6.40. Each of these dissociation constants are associated with a specific proton on the molecule. The influence pH on the distribution of citric acid species at different degrees of deprotonation is shown in Figure 2.2.

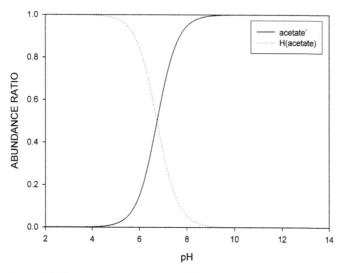

Figure 2.1 Distribution of protonated and deprotonated acetic acid species as a function of pH.

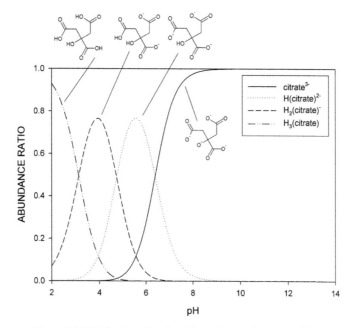

Figure 2.2 Distribution of citric acid species as a function of pH.

2.1.2 The solubility product

The dissolution of a solid ionic compound comprising a cation A^{y+} and an anion B^{x-} in water can be described by the equation:

$$A_xB_y \rightleftharpoons xA^{y+} + yB^{x-}$$

In a case where a compound dissolves to form a solution approaching infinite dilution, the extent to which this reaction occurs is defined by the solubility of the compound which can be described using its solubility product (K_{sp}):

$$K_{sp} = [A^{y+}]^x[B^{x-}]^y$$

The nature of this equation means that the units of the solubility product depend on the nature of the reaction. Where x and y are 1, the units will be mol^2/l^2. However, if x were 2 and y 3, the units would be mol^6/l^6. For convenience the solubility product is usually expressed as $log\ K_{sp}$ with a larger value denoting greater solubility.

By way of example, if we consider the compound calcium hydroxide —$Ca(OH)_2$—the dissolution reaction will take the form:

$$Ca(OH)_2 \rightleftharpoons 2OH^- + Ca^{2+}$$

and so the solubility product of calcium hydroxide is:

$$K_{sp} = [Ca][OH]^2.$$

In the case of calcium hydroxide, $log\ K_{sp}$ is –5.05, and so K_{sp} is 8.91 × $10^{-6}\ mol^3/l^3$. Its solubility in mol/l would therefore be the cube root of K^{sp}: 0.02 mol/l.

In the case of hydroxide compounds, the solubility product can be expressed in a different form. This is because the dissolution of such compounds can be expressed as an alternatively arranged equilibrium equation in the form shown here for ferrihydrite:

$$Fe(OH)_3 + 3H^+ \rightleftharpoons Fe^{3+} + 3H_2O$$

An alternative solubility product K^*_{sp} may be used, which takes the form:

$$K^*_{sp} = [Fe^{3+}][H+]^{-3}$$

For this reason it is usually advisable to always provide the equilibrium equation from which the solubility product has been obtained.

In less dilute solutions where ionic species are involved, the use of concentration to define the solubility product becomes less appropriate. This is because ions will electrostatically interact with each other and neighbouring water molecules. As a result, it is necessary to consider concentration in terms of the 'effective' concentration accounting for these interactions. This effective concentration is called the activity (a) of the ion. The activity can be defined as:

$$a_i = \gamma_i \frac{[i]}{[i^\theta]}$$

where
a_i = the activity of ionic species i;
$[i]$ = the concentration of species i (mol/l);
γ_i = the activity coefficient; and
$[i^\theta]$ = the standard molar concentration (1 mol/l).

The standard molar concentration is required to render the activity dimensionless.

Before it is possible to discuss how the activity coefficient is obtained, it is necessary define one other characteristic of a solution—its ionic strength. Ionic strength (i) is a measure of concentration which takes account of the valence of the ions dissolved, with units of concentration. It is defined by the equation:

$$I = \frac{1}{2}\sum [i]z_i^2$$

where z_i is the charge on ionic species i. The ionic strength equation must include all ions present in a solution.

At low ionic strengths ($I < 0.001$ M) the Debye-Hückel equation can be used to obtain a value for the activity coefficient:

$$\log \gamma_i = -Az_i^2 \sqrt{I}$$

where A is a constant which is independent of the ion involved, but which changes with temperature and pressure. At 25°C and 1 atmosphere pressure, its value is 0.5085 [1]. At higher ionic strengths, different forms of the Debye-Hückel equation must be used. Where the ionic strength is in the range $0.001 \leq I < 0.1$, the extended Debye-Hückel equation applies:

$$\log \gamma_i = \frac{-Az_i^2 \sqrt{I}}{1 + Ba_0 \sqrt{I}}$$

where
B = a constant dependent on temperature and pressure; and
a_0 = the theoretical hydrated radius of the ion (Å).

The hydrated radius of an ion is the radius within which water molecules are closely bound to the ion.

Within the range $0.1 \leq I < 0.7$, the Davies equation [2] should be used:

$$\log \gamma_i = \frac{-Az_i^2 \sqrt{I}}{1 + \sqrt{I}} + bI$$

The original form of the Davies equation used a fixed value of *b*, and this value has been revised several times, but is commonly 0.3. Using a fixed value will introduce some inaccuracy at ionic strengths lower than 0.1 and also for ions with valencies of more than 1. This has subsequently been addressed through the development of ion-specific values of b. Where such an approach is adopted, the Davies equation is referred to as the Truesdell-Jones equation [3].

A number of approaches for obtaining activity coefficients for solutions of ionic strength, including the Pitzer ion interaction and the B-dot equation. These are of lesser relevance to the issues under discussion, but of great value when dealing with environments containing solutions containing brine-like concentrations of highly soluble salts.

Log K_{sp} is influenced by crystal size, temperature and pressure. It is often the case that the influence of crystal size, rightly or wrongly, is often ignored. Pressure usually has a relatively minor influence over the solubility product. Temperature, however, has potentially a much greater influence and, in systems where temperature may vary, this effect will usually need to be accommodated in solubility calculations.

The influence of temperature on solubility is the result of Le Chatelier's principle, which states that a change to a system of concentration, volume, temperature or pressure will cause that system to oppose the change. Dissolution reactions can either be exothermic (giving out heat) or endothermic (taking in heat) and this will change the temperature of the fluid in which the reaction is occurring. Thus, in accordance with Le Chatelier's principle, as the temperature of a system increases, the solubility product of a compound which undergoes an exothermic reaction will decline, whilst the solubility of a compound with an endothermic reaction will increase.

The influence of temperature (expressed in Kelvin) is described by the van't Hoff equation:

$$logK_{sp,T_2} - logK_{sp,T_1} = \frac{\Delta H_R^0}{2.303R}\left(\frac{1}{T_1} - \frac{1}{T_2}\right)$$

where K_{sp,T_2} =   the solubility product at a temperature of T_2;
K_{sp,T_1} =   the solubility product at a reference temperature T_1;
ΔH_R^0 =   the standard enthalpy of reaction (J/mol); and
R =   the gas constant (J/K mol).

In the context of aqueous chemistry, the reference temperature is frequently 25°C. ΔH_R^0 is dependent on temperature, but where temperatures are close to the reference temperature ($\pm$ 20 °K), the value for the reference temperature may be assumed with little compromise in the accuracy of the predicted change in *log* K_{sp}.

The solubility of a substance is influenced by the concentration of other substances dissolved in the solution, where ions present in the substance are already present in solution. If we return to calcium hydroxide, this effect can be understood if we consider its dissolution in a solution of sodium hydroxide (NaOH). If the concentration of NaOH is 0.001 mol/l, the solubility product will be:

$$K_{sp} = [B][B+0.001]^2.$$

where [B] is the concentration of calcium ions in mol/l deriving from the dissolution of calcium hydroxide, and is therefore also the solubility of calcium hydroxide. Therefore, substituting S for [B] and rearranging gives a quadratic equation:

$$0 = S^2 + 0.001S + 0.000001 - K_{sp}.$$

Since we have already seen that K_{sp} is 8.91×10^{-6} mol³/l³

$$0 = S^2 + 0.001S - 0.00000791.$$

Solving this equation gives a value for the solubility of 2.36×10^{-3} mol/l, meaning that the solubility has reduced by an order of magnitude. This effect is known as the *'common ion effect'*.

It should be noted that K_{sp} is also dependent on the quantity of other ions dissolved in the solution—the 'salt effect'. Generally, an increase in dissolved ions will act to reduce the solubility product.

2.1.3 The formation of metal complexes in solution

When a metal ion dissolves in water in the manner described in the previous section it may simply remain in solution as a cation. However, where other molecules or ions are present in solution, there exists the possibility that the cation will form a metal complex with one or more of these entities, which are referred to as ligands when interacting in this way. A complexation reaction can be written in the form

$$M + L \rightleftharpoons ML$$

The complex ML may not be the only complex which can form. For instance, a second ligand may be involved:

$$ML + L \rightleftharpoons ML_2$$

In a similar manner to the acid dissociation constant, the equilibrium of the first reaction can be described using the equation:

$$K_1 = \frac{[ML]}{[M][L]}$$

where K_1 is said to be a *stability constant*, and is specifically the *association constant* of the complex. Similarly, the equilibrium of the second reaction is described thus:

$$K_2 = \frac{[ML_2]}{[ML][L]}$$

These association constants are described as stepwise constants: they describe the individual steps required to move from M to ML_2. Of course, the reaction could be described in terms of both stepwise reactions occurring together:

$$M + 2L \rightleftharpoons ML$$

In which case the equilibrium of the reaction can be characterized by the *cumulative* constant, β_{12}:

$$\beta_{12} = K_1 K_2 = \frac{[ML]}{[M][L]} \cdot \frac{[ML_2]}{[ML][L]} = \frac{[ML_2]}{[M][L]^2}$$

It is, again, often convenient to express stability constants in terms of log K.

The ionic form that a molecule adopts in solution will define to a large extent what complexes it can form with metals. For instance, calcium will form a complex (of the ML type) with the acetate ion ($C_2H_3O_2^-$), but not with its protonated form. This has significant implications, since it means that the concentration of metal complexes in solution will vary depending on pH. This is shown in Figure 2.3 which shows the distribution of the Ca(acetate) complex only appears in significant quantities once the pH is sufficiently high to yield large concentrations of deprotonated acetic acid (as previously seen in Figure 2.1). Another point to note about this plot is

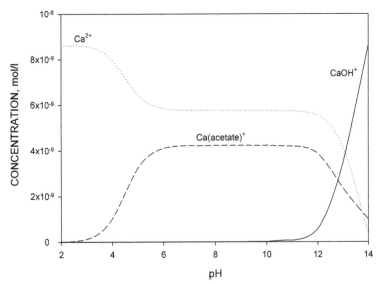

Figure 2.3 Concentrations of complexes formed by calcium in an acetic acid solution as a function of pH. Concentrations: acetic acid: 100 mmol/l; Ca: 0.0001 mmol/l.

that calcium also forms a complex with hydroxide ions (CaOH⁺), whose stability constant is higher than that of Ca(acetate), but whose appearance in significant quantities is observed only once OH⁻ concentrations are high.

As for solubility, stability constants are also influenced by temperature and the effect can again be described by the van't Hoff equation. The salt effect also influences stability constants and, as a result, it is usually the convention to state the ionic strength at which a solubility product was measured.

Particularly stable complexes are formed where the ligand is able to form multiple bonds through different co-ordinating atoms with the same metal ion. This type of interaction is known as chelation, and the stability of the resulting complex is dependent on the size and number of 'chelate rings' formed. A chelate ring is the closed structure formed by the metal ion, a co-ordinating atom in the ligand molecule, through the sequence of bonded atoms that link the co-ordinating atoms, and back to the metal ion. A higher number of chelate rings is likely to increase the stability of the surface complex formed. Moreover, the number of ring members plays an important role in complex stability, with relative stability usually following the sequence:

5-membered > 6-membered > 7-membered/4-membered.

2.1.4 Solubility diagrams and their construction

This chapter describes how some of the substances produced by living organisms interact with the elements present in cement and, to a large extent, concrete. The way in which this has been done is through the use of solubility diagrams. Solubility diagrams are sometimes referred to as predominance plots and define the predominant chemical species in a system for a given pH and concentration of a given species, regardless of whether the species is in solid form or dissolved as an aqueous species. An example of a solubility diagram can be seen in Figure 2.4 for calcium. In the plot, the x-axis is pH, whilst the y-axis is the log of concentration of calcium. The plot shows the predominance of three species: Ca^{2+}, $Ca(OH)^+$ and portlandite ($Ca(OH)_2$). Portlandite is in the solid state, which is indicated by '(s)'. Thus, we are able to see that when the pH is 13 and the concentration of calcium is 0.01 mol/l (i.e., log[Ca] = −2), then portlandite is the predominant species in the system.

It should be stressed that the predominant species will usually not be the only species present. This is a potential weakness of solubility diagrams, since they do not present a comprehensive depiction of the entire system at a given set of conditions. However, as a means of getting an approximate idea of the main changes which occur as, for instance, pH changes in a system, these diagrams are an extremely useful tool.

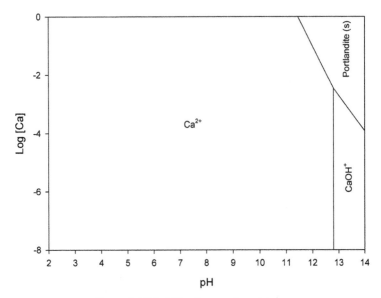

Figure 2.4 Solubility diagram for calcium.

The solubility diagrams used in this chapter have been drawn using the computer program MEDUSA which uses the stability constants and solubility products defined for a system. These constants have been obtained from various sources in the literature. Unless otherwise stated they correspond to a temperature of 25°C. In the case of stability constants, wherever possible the values for an ionic strength of 0 have been selected. Where the effect of acidic species is investigated, all solubility diagrams have been constructed for an acid concentration of 0.1 mol/l.

2.2 Cementitious Systems in the Presence of Water

Before the influence of different acids involved in biodeterioration processes can be discussed, it is necessary to examine the nature of cementitious systems under more conventional chemical conditions. First, the chemical nature of hardened cement paste will be briefly examined, before the manner in which changes in pH affect their constituents.

2.2.1 Introduction to cement chemistry

The clinker that is used to make Portland cement consists of four main phases: tricalcium silicate ($3CaO.SiO_2$), dicalcium silicate ($2CaO.SiO_2$), tricalcium aluminate ($3CaO.Al_2O_3$) and tetracalcium aluminoferrite ($4CaO.Al_2O_3.Fe_2O_3$). This clinker is ground with a quantity of calcium sulfate— usually either as gypsum ($CaSO_4.2H_2O$) or anhydrite ($CaSO_4$).

Portland cement sets and hardens as the result of a series of hydration reactions with water. These reactions occur at notably different rates and the nature of each reaction is influenced to some extent by the other reactions occurring simultaneously. Whilst the details of these reactions do not require examination in any great detail, the products of the reactions do.

The most important product of Portland cement hydration is the poorly crystalline calcium silicate hydrate (CSH) gel. This product has a composition which is able to vary quite considerably. In particular, its Ca/Si ratio can vary from between 0.6 and 1.7 under normal conditions [4]. In addition, other elements can be incorporated into its structure.

Another important constituent is calcium hydroxide (portlandite, $Ca(OH)_2$). This compound acts as the main source of hydroxide ions, and consequently the high pH of water in the pores of hydrated Portland cement. It should be noted, however, that the solubility of portlandite is relatively low. Thus, the high pH is achieved through the release of OH^- ions from portlandite to balance the charge of potassium and sodium ions in solution which were originally present in the clinker as sulfate salts.

Additionally there are two groups of compounds containing aluminium and iron known as the AFt ('aluminoferrite-tri') and AFm (aluminoferrite-mono) phases. The most commonly encountered AFt phase in plain hydrated Portland cement is ettringite, which has the formula $3CaO.(Al,Fe)_2O_3(CaSO_4)_3.32H_2O$. The most commonly encountered AFm phase is monosulphate ($3CaO.(Al,Fe)_2O_3.CaSO_4.12H_2O$). These phases contain sulphate, which derives from the calcium sulfate added to the clinker. However, under different chemical conditions the sulphate can be replaced with a wide range of different anions, and the quantity of chemically combined water can also vary.

It has become increasingly rare for Portland cement to be used as the sole cementitious constituent in concrete. Instead Portland cement is combined with other materials, most of them by-products from industrial processes. The most common of these materials are ground granulated blastfurnace slag (GGBS), fly ash (FA) and silica fume (SF). As will be seen in later chapters, these cement components can sometimes be used to impart greater resistance to acids.

GGBS is a byproduct of iron manufacture and has a chemical composition not entirely unlike Portland cement, but slightly enriched with regards to silica and aluminate. The high pH conditions encountered when used in combination with Portland cement causes it to undergo a latent hydraulic reaction which produces products similar to those formed by the Portland cement. The composition of the slag means that the Ca/Si ratio of the CSH gel formed will be lower, and that larger quantities of aluminoferrite phases will be formed.

FA is ash remaining from the combustion of pulverized coal during electrical power generation and can be encountered in both calcareous (higher CaO) and siliceous (low levels of CaO) forms. Both forms are mainly composed of SiO_2 and Al_2O_3. Whilst some of the minerals found in FA (notably quartz, mullite, hematite and magnetite) are of very low solubility, and react to a limited extent, a large proportion of the material consists of a glassy substance which reacts with portlandite formed by Portland cement to form CSH and aluminoferrites. This reaction is known as a pozzolanic reaction, and the high SiO_2 and Al_2O_3 content of the material mean that, again, the Ca/Si ratio of the CSH gel is reduced and more aluminoferrite phases are formed.

Silica fume is a by-product of the manufacture of silicon for the electronics sector. It is usually almost 100% SiO_2 and undergoes a pozzolanic reaction similar to FA, leading to a modification in the composition of CSH.

The use of all of these materials in combination with Portland cement will also lead to a reduction in the quantity of portlandite formed, since the Portland cement is diluted, and may be reduced further if the material in question undergoes a pozzolanic reaction. As a result, the pH of the pore solutions of hardened cement pastes containing GGBS, FA and SF will be lower than that for Portland cement alone.

Another type of cement which is gaining popularity in many parts of the world are the calcium aluminate cements. These cements contain mainly Al_2O_3 and CaO, although the proportions vary from product to product, unlike Portland cement whose composition has gradually evolved to a relatively uniform composition globally. After hydration, hardened calcium aluminate cements consist largely of $CaO.Al_2O_3.10H_2O$ and $2CaO.Al_2O_3.8H_2O$, with some $Ca_3Al_2O_9.6H_2O$ and amorphous $Al(OH)_3$. The cement will gradually undergo a process of conversion whereby $CaO.Al_2O_3.10H_2O$ and $2CaO.Al_2O_3.8H_2O$ decompose to the last two products. Conversion has been a cause of concern in the past, since the conversion leads to a loss in volume of the cement, leading to the formation of porosity and a loss in strength. However, better understanding of the material means that design processes can take into account conversion. Moreover, as will be seen in later chapters calcium aluminate cements can potentially play a role in achieving resistance to certain forms of biodeterioration.

A variant of calcium aluminate cements are the calcium sulfoaluminate cements. These cements also contain a quantity of sulfate which means that the hydration products formed are usually ettringite, monosulfate and amorphous $Al(OH)_3$ [5].

2.2.2 Redox potential

A number of chemical elements are said to display 'redox' behavior—they are capable of existing in more than one oxidation state. These elements include carbon (which can exist in 4+, 0, 2– and 4– states in natural waters and minerals), hydrogen (1+, 0), iron (3+, 2+), manganese (4+, 3+, 2+), oxygen (0, 2–), nitrogen (3–, 0, 3+, 5+) and sulphur (6+, 4+, 0, 1–, 2–). A number of other trace metals also display redox behavior.

The transition from a high oxidation state (e.g., Fe^{3+}) to a lower one (Fe^{2+})—reduction—requires the species to accept electrons (e^-). This process can be summarized by the reaction:

$$Fe^{3+} + e^- \rightleftharpoons Fe^{2+}$$

The Fe^{3+} here acts as an oxidizing agent—it accepts electrons. This reaction is referred to as a redox couple, but is not a complete chemical reaction, since there is no explanation of the origin of the electron. The electron must come from a reducing agent. In aqueous solution the reducing agent may be water:

$$H^+ + \tfrac{1}{4}O_2 + e^- \rightleftharpoons \tfrac{1}{2}H_2O$$

This is also a redox couple, and combining the two gives:

$$Fe^{3+} + \tfrac{1}{2}H_2O \rightleftharpoons Fe^{2+} + H^+ + \tfrac{1}{4}O_2$$

This is now a complete reaction, with Fe^{3+} and O_2 acting as oxidizing agents and H_2O and Fe^{2+} acting as reducing agents.

The equilibrium state of a redox couple gives a useful measure of whether a system is oxidizing or reducing. Since redox reactions involve transfer of electrons, it is common for this equilibrium to be expressed in terms of the theoretical voltage associated with the reaction. This voltage is known as the redox potential (E_h). If we consider a solution containing two oxidized species (A and B) undergoing reduction to produce two reduced species (C and D):

$$aA + bB + ne^- \rightleftharpoons cC + dD$$

E_h can be determined using the equation:

$$E_h = E° + \frac{RT}{nF} \ln \frac{[A]^a[B]^b}{[C]^c[D]^d}$$

where $E°$ =   the standard potential of the reaction (V);
   R =   the gas constant (8.3145 J/molK);
   T =   temperature (K); and
   F =   the Faraday constant (96,485 J/V g eq).

It is also common for the redox potential to be expressed in terms of the negative logarithm of the electron concentration (pE):

$$pE = -\log_{10}[e^-] = \frac{E_h}{0.05916}$$

The redox potential of a solution is dependent on what redox couples are present, and their concentration. Each redox couple has a characteristic E_h and the redox potential of the solution will reflect the predominant couple.

If the solution experiences conditions which will alter the E_h—for instance, the introduction of a reducing agent—the concentration of the predominant couple will determine the solution's ability to resist this change, with a high concentration providing greater resistance. Where there is a large concentration of a particular redox pair, the solution is said to have a high redox capacity, or to be *'well-poised'*. Once the capacity of the predominant pair is exceeded, the redox potential will shift with relative ease to that of the redox couple with the nearest E_h.

The redox potential can play an important role in determining the oxidation state of elements present in smaller quantities. For instance, if chromium is present in small quantities alongside iron in larger quantities, the redox reaction

$$Fe^{2+} + Cr^{3+} \rightleftharpoons Fe^{3+} + Cr^{6+}$$

will be established. Thus, the relative proportions of Fe^{2+} and Fe^{3+} will dictate the relative proportions of Cr^{3+} and Cr^{6+}. This, in turn, has significant implications for solution chemistry, since—as is the case for chromium—the oxidation state of an element will often influence its solubility.

Redox potential is also influenced by pH, with the redox potential rising with increasing pH.

In Portland cement, there are usually only a small number of dissolved redox pairs present, in relatively small concentrations. These couples include O_2/H_2O, Fe^{3+}/Fe^{2+}, SO_4^{2-}/SO_3^{2-}.

This means that the pore solution within Portland cement is poorly-poised. The redox potential of Portland cement is typically around Eh = +100 – +200 mV. However, where quantities of sulfides are present—for instance, if ground granulated blastfurnace slag has been used in conjunction with Portland cement—the redox potential may be much lower: –400 mV [6]. In constructing solubility plots in this chapter, an Eh of +150 mV was used.

2.2.3 Calcium in water

Calcium makes up a considerable proportion of Portland cement, and so its behaviour in solution plays a very important role in the deterioration of concrete under conditions of changing pH.

In hydrated Portland cement which has not undergone any chemical interactions with its surrounding environment, calcium will be present as CSH gel, portlandite and calcium aluminate (and ferrite) hydrates. For the purposes of summarizing the effect of water on the integrity of hardened Portland cement, the solubility diagram for calcium alone is initially sufficient. In such a case, the only complex which may be formed is $CaOH^+$, whose stability constant is provided in Table 2.1. The only solid phase of relevance is portlandite, whose solubility product is provided in Table 2.2. The solubility diagram constructed using this information is shown in Figure

Table 2.1 Stability constants of complexes formed by calcium, aluminium and iron(II) and (III) ions in water.

Complex	Reaction	Stability Constant	Reference
Ca			
$CaOH^+$	$Ca^{2+} + H_2O \rightarrow CaOH^+ + H^+$	−12.697	[7]
Al			
$AlOH^{2+}$	$Al^{3+} + H_2O \rightleftharpoons AlOH^{2+} + H^+$	−4.997	[7]
$Al(OH)_2^+$	$Al^{3+} + 2H_2O \rightleftharpoons Al(OH)_2^+ + 2H^+$	−10.094	[7]
$Al(OH)_3$	$Al^{3+} + 3H_2O \rightleftharpoons Al(OH)_3 + 3H^+$	−16.791	[7]
$Al(OH)_4^-$	$Al^{3+} + 4H_2O \rightleftharpoons Al(OH)_4^- + 4H^+$	−22.688	[7]
Fe(II)			
$FeOH^+$	$Fe^{2+} + H_2O \rightleftharpoons FeOH^+ + H^+$	−9.397	[7]
$Fe(OH)_2$	$Fe^{2+} + 2H_2O \rightleftharpoons Fe(OH)_2 + 2H^+$	−20.494	[7]
$Fe(OH)_3$	$Fe^{2+} + 3H_2O \rightleftharpoons Fe(OH)_3^- + 3H^+$	−28.991	[7]
Fe(III)			
$FeOH^{2+}$	$Fe^{3+} + H_2O \rightleftharpoons FeOH^{2+} + H^+$	−2.187	[7]
$Fe(OH)_2^+$	$Fe^{3+} + 2H_2O \rightleftharpoons Fe(OH)_2^+ + 2H^+$	−4.594	[7]
$Fe(OH)_3$	$Fe^{3+} + 3H_2O \rightleftharpoons Fe(OH)_3 + 3H^+$	−12.56	[7]
$Fe(OH)_4^-$	$Fe^{3+} + 4H_2O \rightleftharpoons Fe(OH)_4^- + 4H^+$	−21.588	[7]
$Fe_2(OH)_2^{4+}$	$2Fe^{3+} + 2H_2O \rightleftharpoons Fe_2(OH)_2^{4+} + 2H^+$	−2.854	[7]
$Fe_3(OH)_4^{5+}$	$3Fe^{3+} + 4H_2O \rightarrow Fe_3(OH)_4^{5+} + 4H^+$	−6.288	[7]

Table 2.2 Solubility and molar volume data for compounds relevant to the interaction of hydrated Portland cement with water.

Compound	Formula/Reaction	Solubility Product, $\log k_{sp}$	Reference	Molar Volume, cm^3/mol	Reference
Portlandite	$Ca(OH)_2$ $Ca(OH)_2 \rightarrow Ca^{2+} + 2OH^-$	−5.05	[9]	32.9	[10]
Gibbsite	$Al(OH)_3$ $Al(OH)_3 + OH^- \rightleftharpoons Al(OH)_4^+$	−1.40	[9]	32.2	[10]
Ferrihydrite	$Fe(OH)_3$ $Fe(OH)_3 + 3H^+ \rightleftharpoons Fe^{3+} + 3H_2O$	3.191	[7]	28.1	[11]

2.4 which covers a pH range of 2 to 14. The concentration of calcium in the diagram varies between 0.00001 to 1000 mmol/l.

At high pH and at relatively high concentrations of calcium, solid portlandite is the dominant form of calcium. At lower concentrations (below the solubility limit of portlandite) $CaOH^+$ is dominant. It is also evident that even a minor drop in pH below that which is typical for Portland cement will lead to the dissolution of portlandite, with the dissolved calcium ion —Ca^{2+}—becoming dominant. Thus, the portlandite in Portland cement is extremely vulnerable under acidic conditions.

Calcium is, of course, not only present as portlandite. Where it is present as calcium aluminate and ferrite hydrates, a similar vulnerability exists, although this is illustrated best in the section relating to sulphuric acid. The situation for CSH gel is a little more complicated, since this will undergo loss of calcium (decalcification), ultimately leaving a silica gel behind.

2.2.4 Aluminium and iron

Aluminium and iron are in many ways very similar elements, and as a result are able to substitute for each other in the structures of many compounds. This is certainly true of the AFm and AFt phases of cement. One of the key distinguishing features is that, whilst both elements can exist in a number of oxidation states, iron is commonly encountered in two—Fe^{2+} and Fe^{3+}— whereas aluminium usually exists in only the Al^{3+} state.

Figure 2.5 shows the solubility diagram for aluminium. A variety of complexes are formed between aluminium and hydroxide ions—as shown in Table 2.1—and most of these are encountered in the diagram. A significant proportion of this diagram is occupied by the mineral gibbsite ($Al(OH)_3$), which only becomes soluble under very high and very low pH conditions. Crystalline gibbsite is, in fact, seldom encountered in hydrated Portland cement, meaning that it cannot be identified using techniques such as X-ray diffraction. Instead, an amorphous precipitate is formed.

The solubility diagram for iron is shown in Figure 2.6. As for aluminium, much of the diagram's area is occupied by a solid phase, in this case ferrihydrite. Ferrihydrite is an oxyhydroxide compound with a variable composition. The official formula ascribed to it by the International Mineral Association is $5Fe_2O_3.9H_2O$, but the water content varies considerably. Because of this, the formula $Fe(OH)_3$ is often used for simplicity. It has been proposed that ferrihydrite may, in fact, be a mixture of multiple phases, although a single phase structure has also been proposed. The reason for this uncertainty is that ferrihydrite is precipitated as nano-scale particles and so appears amorphous when studied using X-ray diffraction.

The persistence of ferrihydrite even at relatively low pH means that concrete and cement undergoing acid attack will, for many acids, develop a

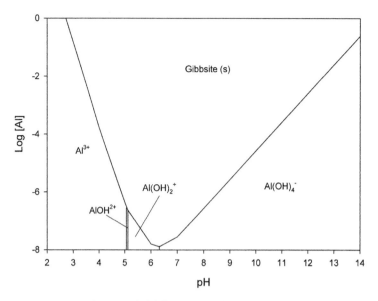

Figure 2.5 Solubility diagram for aluminium.

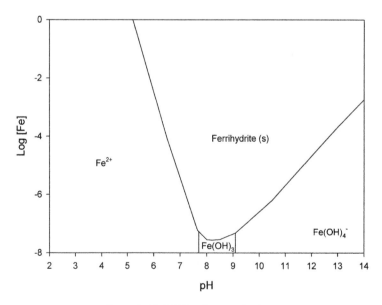

Figure 2.6 Solubility diagram for iron.

red/brown colour, which is sometimes mistakenly interpreted as evidence for steel reinforcement corrosion.

2.3 Inorganic Acids

2.3.1 Sulphuric acid

Sulphuric acid (H_2SO_4) is a diprotic acid (meaning it can lose two protons during dissociation). Its first proton is shed readily (see the extremely low pK_{a1} value in Table 2.3), making it a strong acid. However, the second deprotonation does not occur until a much higher pH is reached. The stability constants of complexes formed between calcium, aluminium and iron are given in Table 2.4.

The interaction of Portland cement with sulphuric acid requires examination both in terms of interaction of the acid separately with calcium, aluminium and iron ions, and with a combination of the two. This is because one potential reaction product is ettringite, $3CaO.(Al_2O_3,Fe_2O_3)$ $(CaSO_4)_3.32H_2O$, whose characteristics relevant to this discussion are shown

Table 2.3 Acid dissociation constants for sulphuric acid.

Acid	Formula	Acid Dissociation Constant		Reference
		pK_{a1}	pK_{a2}	
Sulphuric acid	H_2SO_4	−6.38	1.99	[8, 9]
		HSO_4^-	SO_4^{2-}	
		hydrogen sulphate	sulphate	

Table 2.4 Stability constants of complexes formed by calcium, aluminium and iron(II) and (III) ions in water containing sulphuric acid.

Complex	Reaction	Stability Constant	Reference
Ca			
$CaSO_4$	$Ca^{2+} + SO_4^{2-} \rightleftharpoons CaSO_4$	2.36	
Al			
$AlSO_4^+$	$Al^{3+} + SO_4^{2-} \rightleftharpoons AlSO_4^+$	3.89	
$Al(SO_4)_2^-$	$Al^{3+} + 2SO_4^{2-} \rightleftharpoons Al(SO_4)_2^-$	4.92	[7]
Fe(II)			
$FeSO_4$	$Fe^{2+} + SO_4^{2-} \rightleftharpoons FeSO_4$	2.39	
Fe(III)			
$FeSO_4^+$	$Fe^{3+} + SO_4^{2-} \rightleftharpoons FeSO_4^+$	4.05	
$Fe(SO_4)_2^-$	$Fe^{3+} + 2SO_4^{2-} \rightleftharpoons Fe(SO_4)_2^-$	5.38	

in Table 2.5, along with other solid compounds potentially formed between calcium, aluminium, iron and sulphate ions.

The solubility diagram of the system comprising calcium, aluminium and sulphate is shown in Figure 2.7. At high calcium concentrations the dominant component of the system is portlandite. As the pH falls, portlandite is replaced by solid gypsum ($CaSO_4.2H_2O$). At calcium concentrations below the solubility limit of gypsum, $CaSO_4$ in aqueous solution is predominant, except at high pH, where ettringite is stable. In the absence of aluminium, the solubility diagram essentially remains the

Table 2.5 Solubility and molar volume data for compounds relevant to the interaction of hydrated Portland cement with sulphuric acid.

Compound	Formula/Reaction	Solubility Product, $\log k_{sp}$	Reference	Molar Volume, cm^3/mol	Reference
Gypsum	$CaSO_4.2H_2O \rightleftharpoons Ca^{2+} + SO_4^{2-} + 2H_2O$	−4.43	[10]	74.5	[11]
Ettringite	$Ca_6[Al(OH)_6]_2(SO_4)_3.26H_2O \rightleftharpoons 6Ca^{2+} + 2Al(OH)_4^- + 3SO_4^{-2} + 26H_2O + 4OH^-$	−44.6	[10]	703.6	[11]
Ettringite (Fe)	$Ca_6[Fe(OH)_6]_2(SO_4)_3.26H_2O \rightleftharpoons 6Ca^{2+} + 2Fe(OH)_4^- + 3SO_4^{2-} + 26H_2O + 4OH^-$	−44.0	[13]	717.4	[11]
Monosulphate	$3CaO \cdot Al_2O_3 \cdot CaSO_4 \cdot 12H_2O \rightleftharpoons 4Ca^{2+} + 2Al(OH)_4^- + SO_4^{2-} + 4OH^- + 6H_2O$	−29.4	[10]	308.9	[11]
Monosulphate (Fe)	$3CaO \cdot Fe_2O_3 \cdot CaSO_4 \cdot 12H_2O \rightleftharpoons 4Ca^{2+} + 2Fe(OH)_4^- + SO_4^{2-} + 4OH^- + 6H_2O$	−33.2	[13]	321.3	[11]
Hydrogarnet	$3CaO.Al_2O_3.6H_2O \rightleftharpoons 3Ca^{2+} 2Al(OH)_4^- + 4OH^-$	−22.5	[10]	149.5	[11]
Hydrogarnet (Fe)	$3CaO.Fe_2O_3.6H_2O \rightleftharpoons 3Ca^{2+} + 2Fe(OH)_4^- + 4OH^-$	(≥) −26.3	[14]	155.2	[11]
Basaluminite	$Al_4(OH)_{10}SO_4.5H_2O \rightleftharpoons 4Al^{3+} + SO_4^{2-} + 10OH^- + 5H_2O$	−117.3	[15]	218.9	[12]
Jurbanite	$Al(OH)SO_4.5H_2O \rightleftharpoons Al^{3+} + SO_4^{2-} + OH^- + 5H_2O$	−17.23	[16]	128.9	[12]

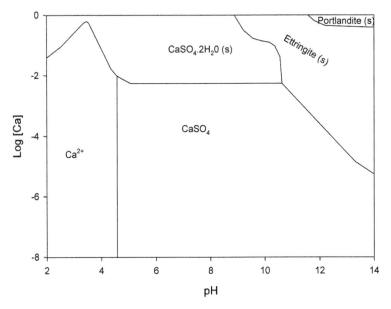

Figure 2.7 Solubility diagram for Ca and Al in the presence of sulphuric acid, with respect to calcium-bearing species and phases. The concentrations of Al and SO_4^{2-} are both fixed at 100 mmol/l.

same, with the exception that ettringite and monosulphate are, of course, absent. Exchanging iron for aluminium, the solubility diagram retains a similar form (Figure 2.8).

Figure 2.9 examines the Ca-Al-SO_4-H_2O system from the perspective of the predominant forms of Al. Ettringite is, again, evident under high pH conditions. As for the case when no sulphate is present, gibbsite is dominant at intermediate pH values, whilst the solid phase Jurbanite $(Al(OH)SO_4.5H_2O)$ is, in theory, precipitated under more acidic conditions. However, in reality, the formation of this phase has not been reported. It has been proposed that in natural soils, a mixture of amorphous $Al(OH)_3$ and basaluminite $(Al_4(OH)_{10}SO_4.5H_2O)$ is actually formed [17]. Whether this is the case for hydrated Portland cement is not known, and it is feasible that Al^{3+} and SO_4^{2-} ions are, in fact, present within the silica gel remaining from the decalcification of CSH gel.

In the Ca-Fe-SO_4-H_2O system (Figure 2.10), the dominance of ferrihydrite is somewhat reduced relative to the situation where sulphate is absent. This is due to the formation of $FeSO_4$ in solution. At very low pH, and in the absence of oxygen, at least in theory, solid iron (II) sulfide forms.

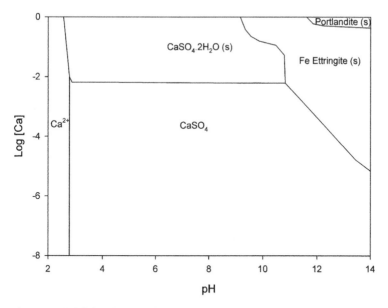

Figure 2.8 Solubility diagram for Ca and Fe in the presence of sulphuric acid, with respect to calcium-bearing species and phases. The concentrations of Fe and SO_4^{2-} are both fixed at 100 mmol/l.

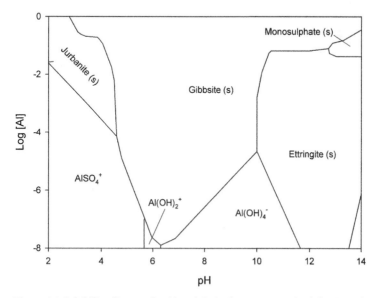

Figure 2.9 Solubility diagram for Al and Ca in the presence of sulphuric acid, with respect to aluminium-bearing species and phases. The concentrations of Ca and SO_4^{2-} are both fixed at 100 mmol/l.

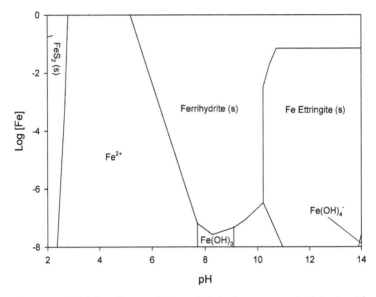

Figure 2.10 Solubility diagram for Fe and Ca in the presence of sulphuric acid, with respect to iron-bearing species and phases. The concentrations of Ca and SO_4^{2-} are both fixed at 100 mmol/l.

2.3.2 Nitric acid

Nitric acid is a strong, monoprotic acid with the formula HNO_3. Table 2.6 gives the dissociation constant of the compound, which dissociates to give the nitrate ion—NO_3^-. Table 2.7 gives the stability constants of complexes formed by the nitrate ion with various elements relevant to cement chemistry. Note that it is possible for the nitrate ion to be reduced to the nitrite ion (NO_2^-) or the ammonium ion (NH_4^+) under the appropriate conditions.

The solubility products of the nitrate salts of calcium, aluminium and iron are shown in Table 2.8. It is evident that these compounds are all very soluble. This, along with the fact that nitric acid is only weakly complexing—indicated by the relatively low stability constants of complexes formed between iron and calcium ions, and the absence of a complex with aluminium—means that it is essentially unnecessary to present solubility diagrams for nitric acid in combination with single metal ions, since the more generic diagrams in Figures 2.4–2.6 suffice.

However, when calcium and aluminium are present together, a nitrate AFm phase may be precipitated. The solubility diagram for the calcium/aluminium/nitrate system is shown in Figure 2.11, showing that the formation of this phase has the effect of limiting the solubilisation of calcium until relatively low pH values. No evidence for an iron analogue

Table 2.6 Acid dissociation constants for nitric acid.

Acid	Formula	Acid Dissociation Constant	Reference
		pK_a	
Nitric acid	HNO_3	-1.3	[18]
		NO_3^-	
		nitrate	

Table 2.7 Stability constants of complexes formed by calcium and iron (III) ions in water containing nitric acid.

Complex	Reaction	Stability Constant	Reference
H⁺			
NO_2^-	$NO_3^- + 2H^+ + 2e^- \rightleftharpoons NO_2^- + H_2O$	28.57	
NH_4^+	$NO_3^- + 10H^+ + 8e^- \rightleftharpoons NH_4^+ + 3H_2O$	119.077	[7]
Ca			
$CaNO_3^+$	$Ca^{+2} + NO_3^- \rightleftharpoons CaNO_3^+$	0.50	
Fe(III)			
$FeNO_3^{+2}$	$Fe^{+3} + NO_3^- \rightleftharpoons FeNO_3^{+2}$	1.00	

Table 2.8 Solubility and molar volume data for compounds relevant to the interaction of hydrated Portland cement with nitric acid.

Compound	Formula/Reaction	Solubility Product, $\log k_{sp}$	Reference	Molar Volume, cm³/mol	Reference
Calcium nitrate tetrahydrate	$Ca(NO_3)_2 \cdot 4H_2O \rightleftharpoons Ca^{2+} + 2NO_3^- + 4H_2O$	3.30	[8]	124.3	[23]
Aluminium nitrate nonahydrate	$Al(NO_3)_3 \cdot 9H_2O \rightleftharpoons Al^{3+} + 3NO_3^- + 9H_2O$	2.45	[19]	220.9	[24]
Nitrate AFm	$3CaO \cdot Al_2O_3 \cdot Ca(NO_3)_2 \cdot 10H_2O \rightleftharpoons 4Ca^{2+} + 2Al(OH)_4^- + 2NO_3^- + 4OH^- + 4H_2O$	-28.67	[20]	148.3	[21]
Iron (II) nitrate hexahydrate	$Fe(NO_3)_2 \cdot 6H_2O \rightleftharpoons Fe^{2+} + 2NO_3^- + 6H_2O$	2.86	[22]	unknown	–
Iron (III) nitrate hexahydrate	$Fe(NO_3)_3 \cdot 6H_2O \rightleftharpoons Fe^{3+} + 3NO_3^- + 6H_2O$	2.92	[8]	197.6	[25]
Iron (III) nitrate nonahydrate	$Fe(NO_3)_3 \cdot 9H_2O \rightleftharpoons Fe^{3+} + 3NO_3^- + 9H_2O$	3.56	[22]	224.4	[25]

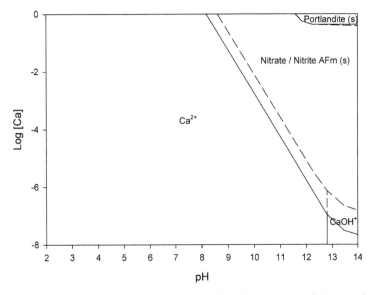

Figure 2.11 Solubility diagram for Ca and Al in the presence of nitric acid and nitrous acid, with respect to calcium-bearing species and phases. The concentrations of Ca and NO_3^-/NO_2^- are all fixed at 100 mmol/l. Solid line = nitric acid; dashed line = nitrous acid.

of the nitrate AFm phase has been found in the literature, but its existence is possible.

2.3.3 Nitrous acid

Nitrous acid is considerably weaker than nitric acid (Table 2.9), dissociating to form nitrite ions. It forms even weaker complexes with calcium than nitric acid, and does not form complexes with aluminium or iron (Table 2.10). In the presence of calcium, nitrite ions can form the salt calcium nitrite monohydrate (Table 2.11), although this compound is relatively soluble. Thus, as for nitric acid, Figures 2.4–2.6 act as solubility diagrams for nitrous acid with single metal ions. A nitrite AFm phase also exists, with similar characteristics to the nitrate AFm phase. A calcium/aluminium/nitrite solubility diagram is also plotted in Figure 2.11, alongside that for nitrate. As for nitrate AFm, it is possible that an iron-bearing form of the nitrite AFm phase exists, but this has yet to be documented.

2.3.4 Carbonic acid and 'aggressive' carbon dioxide

When water is in contact with an atmosphere containing carbon dioxide (CO_2), the gas will dissolve to some degree, depending on its partial pressure

Table 2.9 Acid dissociation constants for nitrous acid.

Acid	Formula	Acid dissociation constant	Reference
		pK_a	
Nitrous acid	HNO_2	3.3	[26]
		NO_3^-	
		nitrite	

Table 2.10 Stability constants of complexes formed by calcium in water containing nitrous acid.

Complex	Reaction	Stability Constant	Reference
Ca			
$CaNO_2^+$	$Ca^{2+} + NO_2^- \rightleftharpoons CaNO_2^+$	-0.28	
$Ca(NO_2)_2$	$Ca^{2+} + 2NO_2^- \rightleftharpoons Ca(NO_2)_2$	-1.03	[27]

Table 2.11 Solubility and molar volume data for compounds relevant to the interaction of hydrated Portland cement with nitrous acid.

Compound	Formula/Reaction	Solubility Product, $\log k_{sp}$	Reference	Molar Volume, cm^3/mol	Reference
Calcium nitrite tetrahydrate	$Ca(NO_2)_2 \cdot 4H_2O \rightleftharpoons Ca^{2+} + 2NO_2^- + 4H_2O$	2.30	[22]	67.3	[22]
Nitrite AFm	$3CaO \cdot Al_2O_3 \cdot Ca(NO_2)_2 \cdot 10H_2O \rightleftharpoons 4Ca^{2+} + 2Al(OH)_4^- + 2NO_2^- + 4OH^- + 4H_2O$	-26.87	[20]	unknown	

and the temperature. This dissolved CO_2 is often referred to as carbonic acid (H_2CO_3), although in reality it will be dissociated to bicarbonate or carbonate ions to a degree determined by pH and temperature. The dissociation constants for carbonic acid at 20°C are given in Table 2.12.

The relationship between the concentration of dissolved carbonic acid —$[H_2CO_3]$—and the partial pressure of CO_2–P_{CO_2}—is given by the equation:

$$[H_2CO_3] = K_{CO_2} P_{CO_2}$$

where K_{CO_2} is an equilibrium constant whose value is $1 \times 10^{1.41}$ at 20°C, and whose magnitude increases with temperature [28].

Carbonic acid undergoes very little by way of interaction with ions encountered in Portland cement, as shown in Table 2.13. However, its interaction with calcium is of great importance. Under higher pH conditions, calcium carbonate ($CaCO_3$) is formed (Figure 2.12). The most stable form of calcium carbonate at normal ambient conditions is calcite, although vaterite is often also formed, gradually converting to calcite. The aragonite form may be precipitated at elevated temperatures. At lower pH values, calcium forms $CaHCO_3^+$, with the charge balanced by a bicarbonate ion. This means that calcium in this configuration can be viewed as being present as $Ca(HCO_3)_2$ —calcium hydrogencarbonate (also referred to as calcium bicarbonate).

Calcium hydrogencarbonate is essentially a hypothetical entity, since it implies that a salt with this composition can be precipitated from solution, which is not actually the case. However, it is useful in explaining the effect of the presence of calcium in a solution exposed to CO_2. The reaction which converts calcium carbonate to calcium hydrogencarbonate can be written as:

$$H_2CO_3 + CaCO_3 \rightleftharpoons Ca(HCO_3)_2$$

The equilibrium of this reaction is shifted to the right by a drop in pH, such as through the dissolution of more carbonic acid. However, calcium hydrogencarbonate associates two molecules of CO_2 to one calcium ion, compared to $CaCO_3$, where the ratio is 1:1. Therefore, the formation of

Table 2.12 Acid dissociation constants for carbonic acid.

Acid	Formula	Acid Dissociation Constant		Reference
		pK_{a1}	pK_{a2}	
Carbonic acid	H_2CO_3	6.38	10.38	[28]
		HCO_3^-	CO_3^{2-}	
		bicarbonate	carbonate	

Table 2.13 Stability constants of complexes formed by calcium and iron(II) and (III) ions in water containing carbonic acid.

Complex	Reaction	Stability Constant	Reference
Ca			
	$Ca^{+2} + H^+ + CO_3^{2-} \rightleftharpoons CaHCO_3^+$	11.60	
	$Ca^{2+} + CO_3^{2-} \rightleftharpoons CaCO_3$	3.20	[34]
Fe(II)			
	$Fe^{2+} + H^+ \rightarrow CO_3^{2-} \rightleftharpoons FeHCO_3^+$	11.43	

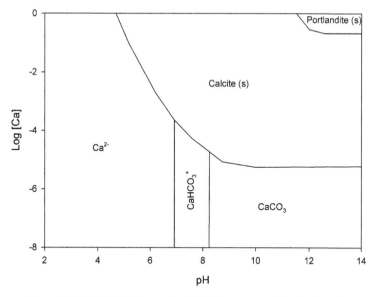

Figure 2.12 Solubility diagram for Ca in the presence of carbonic acid.

calcium hydrogencarbonate acts to limit the decrease in pH as more CO_2 dissolves into water.

This effect is shown in Figure 2.13, which plots the pH of two quantities of water exposed to increasing partial pressures of CO_2. Where only water is present there is rapid decrease in pH as partial pressure increases, with pH levelling out at higher partial pressures. This behavior is to be expected given the logarithmic nature of pH. However, when a small quantity of portlandite is introduced into the water, the shape of the curve is altered, with higher pH values persisting to greater partial pressures. Thus, in systems containing Portland cement, dissolution of the cement matrix will only occur at relatively high CO_2 concentrations where enough is present to exceed the capacity for calcium hydrogencarbonate to act as a 'sink'. The resulting 'free' CO_2 is sometimes referred to as 'aggressive CO_2'.

Table 2.14 provides solubility products for calcium and iron carbonates, but also includes calcium aluminate and ferrite phases which contain carbonate ions. These include the AFt carbonate phase—which is structurally analogous to ettringite—and monocarbonate and hemicarbonate phases, which are members of the AFm group of cement hydration products along with monosulfate. Figures 2.14 and 2.15 show solublity diagrams for systems in which Ca and Al are present. The dominant calcium aluminate phase in both diagrams is monocarbonate, although it should be noted that, overall, the solubility diagrams are not altered much.

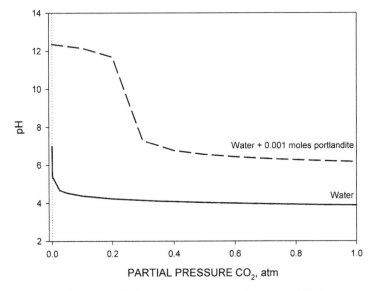

Figure 2.13 Change in pH of water versus the partial pressure of CO_2 in contact with it for two scenarios: (i) pure water; (ii) water in contact with portlandite.

Figure 2.16 is a solubility diagram for iron in the presence of carbonic acid. The main features of this plot that differentiate it from the diagram obtained in the absence of carbonic acid is the presence of the $FeHCO_3^+$ complex and the mineral siderite ($FeCO_3$). Figures 2.17 and 2.18 show the situation where both Ca and Fe are present. As for Al, the presence of the iron analogue of monocarbonate is the only significant change to the diagrams.

2.4 Organic Acids

2.4.1 Formic acid

Formic acid is a carboxylic acid with the formula HCOOH, making it the simplest of carboxylic acid compounds. It is monoprotic and is relatively weak (Table 2.15). It is capable of forming complexes with Ca, Al and Fe(III) ions (Table 2.16). Although the acid mainly forms weak complexes, $Al_3(OH)_2(CHO_2)_6^+$ is the exception to this. Formic acid forms salts with Ca, Al and Fe(II), all of which are relatively soluble in water (Table 2.17). Aluminium formate and iron (III) formate have previously been thought to have the formulae $Al(CHO_2)_3 \cdot 3H_2O$ and $Fe(CHO_2)_3 \cdot H_2O$ respectively. However, a recent crystallographic study has concluded that the compounds have the formulas $Al(CHO_2)_3(CH_2O_2)_{0.25}(CO_2)_{0.75} \cdot 0.25H_2O$ and $Fe(CHO_2)_3(CH_2O_2)_{0.25}$ $(CO_2)_{0.75} \cdot 0.25H_2O$, where CH_2O_2 is a fully protonated formic acid molecule

Table 2.14 Solubility and molar volume data for compounds relevant to the interaction of hydrated Portland cement with carbonic acid.

Compound	Formula/Reaction	Solubility Product, log k_{sp}	Reference	Molar Volume, cm³/mol	Reference
Vaterite	$CaCO_3 \rightleftharpoons Ca^{2+} + CO_3^{2-}$	−7.91	[29]	38	[11]
Aragonite	$CaCO_3 \rightleftharpoons Ca^{2+} + CO_3^{2-}$	−8.30	[7]	34	[11]
Calcite	$CaCO_3 \rightleftharpoons Ca^{2+} + CO_3^{2-}$	−8.48	[7]	37	[11]
Siderite	$FeCO_3 \rightleftharpoons Fe^{2+} + CO_3^{2-}$	−10.24	[7]	29	[30]
AFt carbonate	$Ca_6Al_2(CO_3)_3(OH)_{12}.24H_2O$ $\rightleftharpoons 6Ca^{2+} + 2Al(OH)_4^- + 3CO_3^{2-} + 4OH^- + 24H_2O$	−49.19	[31]	652	[11]
Monocarbonate	$Ca_4Al_2(CO_3)(OH)_{12}.5H_2O \rightleftharpoons$ $4Ca^{2+} + 2Al(OH)_4^- + CO_3^{2-} + 4OH^- + 5H_2O$	−31.47	[32]	262	[32]
Monocarbonate (Fe)	$Ca_4Fe_2(CO_3)(OH)_{12}.6H_2O \rightleftharpoons$ $4Ca^{2+} + 2Fe(OH)_4^- + CO_3^{2-} + 4OH^- + 6H_2O$	−34.59	[32]	292	[32]
Hemicarbonate	$Ca_4Al_2(CO_3)_{0.5}(OH)_{13}.6H_2O \rightleftharpoons$ $4Ca^{2+} + 2Al(OH)_4^- + 0.5CO_3^{2-} + 5OH^- + 6H_2O$	−29.13	[32]	285	[32]
Hemicarbonate (Fe)	$Ca_4Fe_2(CO_3)_{0.5}(OH)_{13}.4H_2O \rightleftharpoons$ $4Ca^{2+} + 2Fe(OH)_4^- + 0.5CO_3^{2-} + 5OH^- + 4H_2O$	−30.83	[32]	273	[32]
Thaumasite	$Ca_6Si_2(SO_4)_2(CO_3)_2(OH)_{12}.24H_2O$ $\rightleftharpoons 2H_3SiO_4^- + 6Ca^{2+} + 2SO_4^{2-} + 2CO_3^{2-} + 2OH^- + 26H2O$	−49.40	[31]	331	[11]

and CO_2 is a gaseous carbon dioxide molecule trapped in the cage-like structures of the compounds.

AFt and AFm phases containing the formate ion have also been reported and characterized in terms of their structure [38]. Whilst solubility data for these phases is not available, it is likely that behaviour comparable to that seen for sulphuric acid is probable.

The solubility diagram for calcium (Figure 2.19) includes a large area in which the 1:1 complex between calcium and the formate ion dominates. However, from a solubility perspective, this has little influence over the areas

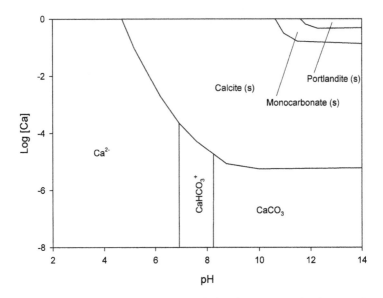

Figure 2.14 Solubility diagram for Ca and Al in the presence of carbonic acid, with respect to calcium-bearing species and phases. The concentrations of Al and CO_3^{2-} are fixed at 100 mmol/l.

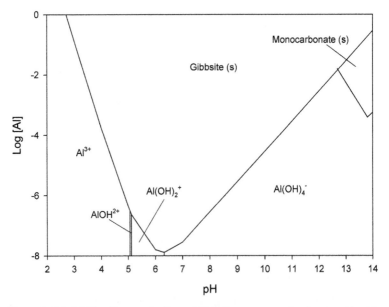

Figure 2.15 Solubility diagram for Ca and Al in the presence of carbonic acid, with respect to aluminium-bearing species and phases. The concentrations of Ca and CO_3^{2-} are fixed at 100 mmol/l.

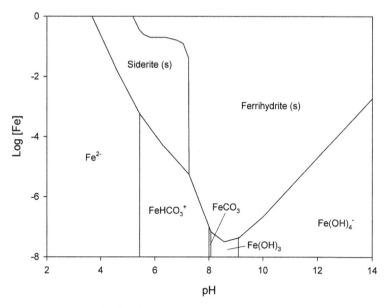

Figure 2.16 Solubility diagram for Fe in the presence of carbonic acid.

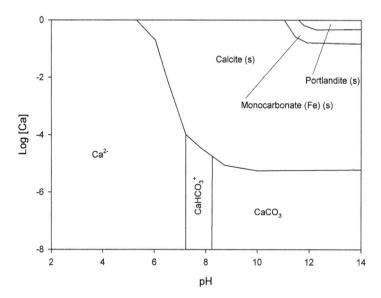

Figure 2.17 Solubility diagram for Ca and Fe in the presence of carbonic acid, with respect to calcium-bearing species and phases. The concentrations of Fe and CO_3^{2-} are fixed at 100 mmol/l.

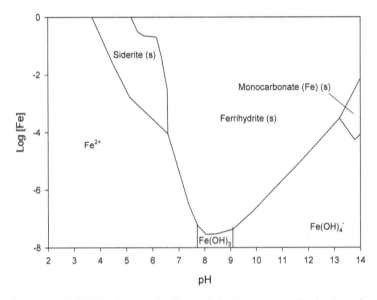

Figure 2.18 Solubility diagram for Ca and Fe in the presence of carbonic acid, with respect to iron-bearing species and phases. The concentrations of Ca and CO_3^{2-} are fixed at 100 mmol/l.

Table 2.15 Acid dissociation constants for formic acid.

Acid	Formula	Acid Dissociation Constant	Reference
		pK_a	
Formic acid	HCOOH	3.75	[33]
		HCO_2^-	
		formate	

Table 2.16 Stability constants of complexes formed by calcium, aluminium and iron (III) ions in water containing formic acid.

Complex	Reaction	Stability Constant	Reference
Ca			
	$Ca^{2+} + CHO_2^- \rightleftharpoons Ca(CHO_2)^+$	1.43	
Al			
	$Al^{3+} + CHO_2^- \rightleftharpoons Al(CHO_2)^{2+}$	1.36	
	$3Al^{3+} + 2OH^- + 6CHO_2^- \rightleftharpoons Al_3(OH)_2(CHO_2)_6^+$	19.90	[33]
Fe(III)			
	$Fe^{3+} + CHO_2^- \rightleftharpoons Fe(CHO_2)^{2+}$	3.10	

Table 2.17 Solubility and molar volume data for compounds relevant to the interaction of hydrated Portland cement with formic acid.

Compound	Formula/Reaction	Solubility Product, $\log k_{sp}$	Reference	Molar Volume, cm^3/mol	Reference
Calcium formate	$Ca(CHO_2)_2 \rightleftharpoons Ca^{2+} + 2CHO_2^-$	0.92	[34]	64	[35]
'Aluminium formate'	$Al(CHO_2)_3(CH_2O_2)_{0.25}(CO_2)_{0.75} \cdot 0.25H_2O \rightleftharpoons Al^{3+} + 3CHO_2^- + 0.75CO_2(g) + 0.25CH_2O_2 + 0.25H_2O$	-0.98	[34]	117	[36]
Aluminium formate hemihydrate (monobasic)	$AlOH(CHO_2)_2 \cdot 0.5H_2O \rightleftharpoons Al^{3+} + OH^- + 2CHO_2^- + 0.5H_2O$	-1.17	[34]	Unknown	–
Formate AFm	$Ca_4Al_2(CHO_2)_2(OH)_{12} \cdot 5H_2O \rightleftharpoons 4Ca^{2+} + 2Al(OH)_4^- + 2CHO_2^- + 4OH^- + 5H_2O$	Unknown	–	Unknown	–
Formate AFt	$Ca_6Al_2(CHO_2)_6(OH)_{12} \cdot 26H_2O \rightleftharpoons 6Ca^{2+} + 2Al(OH)_4^- + 6CHO_2^- + 4OH^- + 26H_2O$	Unknown	–	Unknown	–
Iron (II) formate dihydrate	$Fe(CHO_2)_2 \cdot 2H_2O \rightleftharpoons Fe^{2+} + 2CHO_2^- + 2H_2O$	-1.30	[34]	88	[37]
'Iron (III) formate'	$Fe(CHO_2)_3(CH_2O_2)_{0.25}(CO_2)_{0.75} \cdot 0.25H_2O \rightleftharpoons Fe^{3+} + 3CHO_2^- + 0.75CO_2(g) + 0.25CH_2O_2 + 0.25H_2O$	'soluble'	[8]	126	[36]

in which solid phases dominate. In the case of aluminium, the formation of the $Al_3(OH)_2(CHO_2)_6^+$ complex has the effect of assuming dominance over solid gibbsite in the lower pH region of the plot (Figure 2.20).

2.4.2 Acetic acid

Like formic acid, acetic acid is a carboxylic acid. The acetic acid molecule contains two carbon and is slightly weaker than formic acid (Table 2.18).

The acetate ion forms weak complexes with Ca, Al and Fe(II) ions, but forms stronger complexes with Fe(III) (Table 2.19). Many of the salts formed by acetic acid with Ca, Al and Fe are soluble (Table 2.20). Aluminium

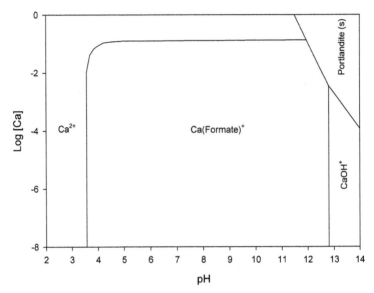

Figure 2.19 Solubility diagram for Ca in the presence of formic acid.

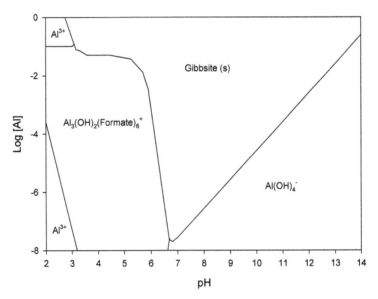

Figure 2.20 Solubility diagram for Al in the presence of formic acid.

Table 2.18 Acid dissociation constants for acetic acid.

Acid	Formula	Acid Dissociation Constant	Reference
		pK_a	
Acetic acid	CH_3COOH	4.756	[8]
		$CH_3CO_2^-$	
		acetate	

Table 2.19 Stability constants of complexes formed by calcium, aluminium and iron(II) and (III) ions in water containing acetic acid.

Complex	Reaction	Stability Constant	Reference
Ca			
	$Ca^{2+} + CH_3CO_2^- \rightleftharpoons Ca(CH_3CO_2)^+$	1.18	
Al			
	$Al^{3+} + CH_3CO_2^- \rightleftharpoons Al(CH_3CO_2)^{2+}$	1.51	[33]
Fe(II)			
	$Fe^{2+} + CH_3CO_2^- \rightleftharpoons Fe(CH_3CO_2)^+$	1.40	
Fe(III)			
	$Fe^{3+} + CH_3CO_2^- \rightleftharpoons Fe(CH_3CO_2)^{2+}$	4.02	
	$Fe^{3+} + 2CH_3CO_2^- \rightleftharpoons Fe(CH_3CO_2)_2^+$	7.57	
	$Fe^{3+} + 3CH_3CO_2^- \rightleftharpoons Fe(CH_3CO_2)_3$	9.59	

diacetate $(Al(CH_3CO_2)_2OH)$ and iron (III) diacetate $(Fe(CH_3CO_2)_2OH)$ are identified as being insoluble in the literature, but quantitative solubility data is currently unavailable. For this reason, in generating the solubility diagrams for Al and Fe, a solubility of 0.05 g/100 g solution—which is generally considered to be the mid-point of the range of solubilities for substances described as being 'practically insoluble'—has been assumed. It should be noted that there also exists an anhydrous calcium acetate salt—$Ca(CH_3CO_2)_2$—although this is of little relevance in the context of aqueous systems, since it reacts with moisture rapidly to give the hydrated form of the salt.

Figure 2.21 shows a solubility diagram for calcium in the presence of acetic acid, with the $Ca(CH_3CO_2)^+$ complex occupying a substantial proportion of the plotted area, but with little change with regards to the solid phases present. The same is true of aluminium (Figure 2.22) with only the presence of the $Al(CH_3CO_2)^{2+}$ complex distinguishing it from the solubility plot obtained in the absence of acetic acid. The acetate ion forms relatively

Table 2.20 Solubility and molar volume data for compounds relevant to the interaction of hydrated Portland cement with acetic acid.

Compound	Formula/Reaction	Solubility Product, log k_{sp}	Reference	Molar Volume, cm³/mol	Reference
Calcium diacetate monohydrate	$Ca(CH_3CO_2)_2.H_2O \rightleftharpoons$ $Ca^{2+} + 2CH_3CO_2^- +$ H_2O	1.38	[39]	117	[40]
Calcium diacetate dihydrate	$Ca(CH_3CO_2)_2.2H_2O \rightleftharpoons$ $Ca^{2+} + 2CH_3CO_2^- +$ $2H_2O$	1.76	[41]	unknown	–
Calcium hydrogen triacetate monohydrate	$CaH(CH_3CO_2)_3.H_2O$ $\rightleftharpoons Ca^{2+} + CH_3COOH +$ $2CH_3CO_2^- + H_2O$	3.07	[41]	158	[42]
Aluminium triacetate	$Al(CH_3CO_2)_3 \rightleftharpoons Al^{3+} +$ $3CH_3CO_2^-$	'Soluble'	[43]	unknown	–
Aluminium diacetate	$Al(CH_3CO_2)_2OH \rightleftharpoons Al^{3+}$ $+ 2CH_3CO_2^- + OH^-$	'Insoluble'	[8]	unknown	–
Aluminium monoacetate	$Al(CH_3CO_2)OH_2 \rightleftharpoons Al^{3+}$ $+ CH_3CO_2^- + 2OH^-$	Assumed to be soluble	–	unknown	–
Iron (II) acetate	$Fe(CH_3CO_2)_2.4H_2O$ $\rightleftharpoons Fe^{2+} + 2CH_3CO_2^- +$ $4H_2O$	'Soluble'	[8]	127	[44]
Iron (III) diacetate	$Fe(CH_3CO_2)_2OH \rightleftharpoons Fe^{3+}$ $+ 2CH_3CO_2^- + OH^-$	'Insoluble'	[8]	unknown	–

stable complexes with Fe (III), but, again, this has a negligible influence over the manner in which solid phases are precipitated (Figure 2.23).

2.4.3 Lactic acid

Lactic acid is a carboxylic acid with the formula $CH_3CH(OH)COOH$. The second carbon atom in the sequence described by the formula is an *asymmetric carbon atom*, meaning that it is attached to four different groups of atoms ($-OH$, $-CH_3$, $-H$ and $-COOH$). Where such a structural feature is present in a molecule, it can be certain that the compound is chiral—its molecules can exist as more than one optical isomer. In the case of lactic acid two isomers exist—the D- and L-forms which are mirror images of each other. It is the L form which is produced by living organisms. Where a mixture of the two forms is present, it is referred to as DL lactic acid. Differences exist between many of the properties of these two forms, and

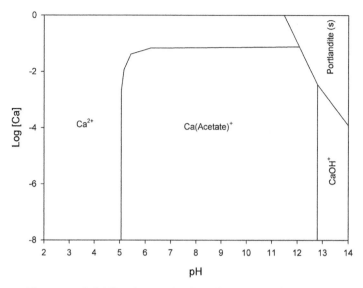

Figure 2.21 Solubility diagram for Ca in the presence of acetic acid.

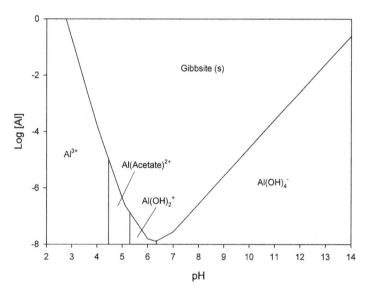

Figure 2.22 Solubility diagram for Al in the presence of acetic acid.

this extends to their behaviour in solution with metal ions. Thus, wherever possible, constants relating to lactic acid have been obtained for the L-form.

Lactic acid contains two –OH groups, but only the carboxylate group normally undergoes deprotonation (Table 2.21). However, interaction with

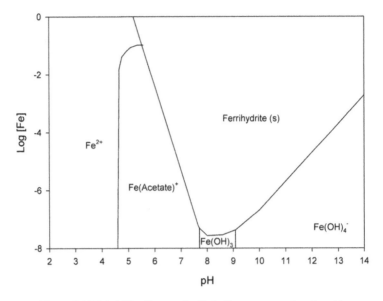

Figure 2.23 Solubility diagram for Fe in the presence of acetic acid.

Table 2.21 Acid dissociation constants for lactic acid.

Acid	Formula	Acid Dissociation Constant	Reference
		pk_a	
Lactic acid	$CH_3CH(OH)COOH$	3.86	[46]
		$CH_3CH(OH)CO_2^-$	
		lactate	

some metal ions allows further deprotonation, as evidenced by the existence of the $Al(CH_3COCO_2)(CH_3C(OH)CO_2)$ complex (Table 2.22). The complexes formed between lactic acid with calcium and aluminium ions are weak. It is likely that iron also forms complexes with lactic acid, but reliable stability constants have not been reported in the literature [45].

Lactic acid forms salts with calcium, aluminium and iron. However, they are all of relatively high solubility (Table 2.23). This, coupled with the nature of the complexes formed, means that the solubility diagrams remain largely unchanged, with the exception of aluminium, where the $Al(CH_3COCO_2)(CH_3C(OH)CO_2)$ complex prevails over gibbsite under more acidic conditions (Figure 2.24).

Table 2.22 Stability constants of complexes formed by calcium and aluminium ions in water containing lactic acid.

Complex	Reaction	Stability Constant	Reference
Ca			
	$Ca^{2+} + CH_3CH(OH)CO_2^- \rightleftharpoons Ca(CH_3CH(OH)CO_2)^+$	0.90	
	$Ca^{2+} + 2CH_3CH(OH)CO_2^- \rightleftharpoons Ca(CH_3CH(OH)CO_2)_2$	1.24	[33]
Al			
	$Al^{3+} + CH_3CH(OH)CO_2^- \rightleftharpoons Al(CH_3CH(OH)CO_2)^{2+}$	1.21	
	$Al^{3+} + 2CH_3CH(OH)CO_2^- \rightleftharpoons Al(CH_3CH(OH)CO_2)_2^+$	2.72	[47]
	$Al^{3+} + 3CH_3CH(OH)CO_2^- \rightleftharpoons Al(CH_3CH(OH)CO_2)_3$	4.92	(DL-Lactate)
	$Al^{3+} + 2CH_3CH(OH)CO_2^- \rightleftharpoons Al(CH_3CHOCO_2)$ $(CH_3CH(OH)CO_2) + H^+$	6.17	

Table 2.23 Solubility and molar volume data for compounds relevant to the interaction of hydrated Portland cement with lactic acid.

Compound	Formula/Reaction	Solubility Product, $\log k_{sp}$	Reference	Molar Volume, cm³/mol	Reference
Calcium lactate pentahydrate	$Ca(CH_3CH(OH)$ $CO_2)_2.5H_2O$ $\rightleftharpoons Ca^{2+} + 2CH_3CH(OH)$ $CO_2^- + 5H_2O$	−0.97	[48]	–	–
Aluminium lactate	$Al(CH_3CH(OH)CO_2)_3$ $\rightleftharpoons Al^{3+} + 3CH_3CH(OH)$ CO_2^-	1.16	[48]	212	[49]
Iron (II) lactate trihydrate	$Fe(CH_3CH(OH)$ $CO_2)_2.3H_2O \rightleftharpoons Fe^{2+} +$ $2CH_3CH(OH)CO_2^- +$ $3H_2O$	−1.87	[48]	–	–
Iron (III) lactate	$Fe(CH_3CH(OH)CO_2)_3 \rightleftharpoons$ $Fe^{3+} + 3CH_3CH(OH)CO_2^-$	'soluble'	[8]	–	–

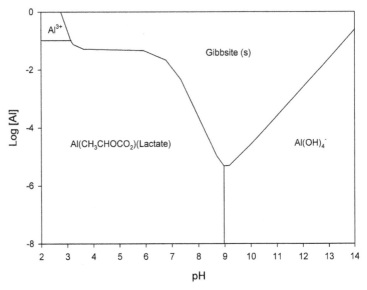

Figure 2.24 Solubility diagram for Al in the presence of lactic acid.

2.4.4 Glycolic acid

Glycolic acid has the formula $CH_2(OH)COOH$ and possesses a carboxylate group as well as a hydroxyl group. Normally, it is only the carboxyl group which undergoes deprotonation to form a glycolate ion (Table 2.24). However, in the presence of iron (III) the hydroxyl group may also deprotonate (Table 2.25).

Very little data exists on the glycolate salts of Ca, Al and Fe both in terms of their solubility and crystal structure (Table 2.26), and no evidence can be found for Fe (II) salts. However, it would appear that all of the compounds formed are relatively soluble. The Ca and Al solubility diagrams in the presence of glycolic acid are shown in Figures 2.25 and 2.26. The diagram for iron is particularly of note, since it shows that ferrihydrite is destabilized by the formation of iron-glycolate complexes.

2.4.5 Oxalic acid

Oxalic acid—$C_2O_4H_2$—is a relatively strong acid that can shed two protons from two carboxylate groups. The acid dissociation constants of oxalic acid are provided in Table 2.27. The stability constants of complexes formed between the acid and Ca and Fe ions are provided in Table 2.28.

Oxalic acid is capable of forming a number of salts with calcium of relatively low solubility, the only difference in composition being the amount of associated water of crystallization. Three of these compounds are

Table 2.24 Acid dissociation constants for glycolic acid.

Acid	Formula	Acid Dissociation Constant pK_a	Reference
Glycolic acid	$CH_2(OH)COOH$	3.83	[33]
		$CH_2(OH)CO_2^-$ glycolate	

Table 2.25 Stability constants of complexes formed by calcium and iron(II) and (III) ions in water containing glycolic acid.

Complex	Reaction	Stability Constant	Reference
Ca			
	$Ca^{2+} + CH_2(OH)CO_2^- \rightleftharpoons Ca(CH_2(OH)CO_2)^+$	1.62	[33]
Fe (II)			
	$Fe^{2+} + CH_2(OH)CO_2^- \rightleftharpoons Fe(CH_2(OH)CO_2)^+$	1.33	[33]
Fe (III)			
	$Fe^{3+} + CH_2(OH)CO_2^- \rightleftharpoons Fe(CH_2(OH)CO_2)^{2+}$	2.90	
	$Fe^{3+} + CH_2(OH)CO_2^- \rightleftharpoons Fe(CH_2OCO_2)^+ + H^+$	4.21	
	$Fe^{3+} + 2CH_2(OH)CO_2^- \rightleftharpoons Fe(CH_2OCO_2)$ $(CH_2(OH)CO_2) + H^+$	6.61	[33]
	$Fe^{3+} + 3CH_2(OH)CO_2^- \rightleftharpoons$ $Fe(CH_2OCO_2)_2(CH_2(OH)CO_2)^{2-} + H^+$	8.11	

Table 2.26 Solubility and molar volume data for compounds relevant to the interaction of hydrated Portland cement with glycolic acid.

Compound	Formula/Reaction	Solubility Product, log k_{sp}	Reference	Molar Volume, cm³/mol	Reference
Calcium glycolate monohydrate	$Ca(CH_2(OH)CO_2)_2 \cdot H_2O \rightleftharpoons$ $Ca^{2+} + 2CH_2(OH)CO_2^- + H_2O$	−2.71	[50]	unknown	−
Aluminium glycolate	$Al(CH_2(OH)CO_2)_3 \rightleftharpoons Al^{3+} +$ $3CH_2(OH)CO_2$	Assumed to be soluble	−	unknown	−
Iron (III) glycolate	$Fe(CH_2(OH)CO_2)_3 \rightleftharpoons Fe^{3+} +$ $3CH_2(OH)CO_2^-$	Assumed to be soluble	−	unknown	−

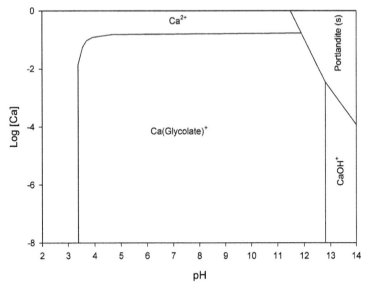

Figure 2.25 Solubility diagram for Ca in the presence of glycolic acid.

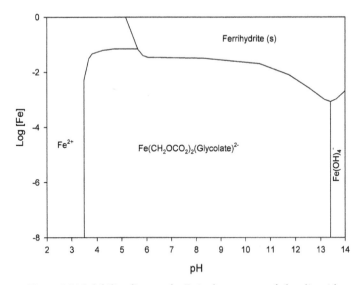

Figure 2.26 Solubility diagram for Fe in the presence of glycolic acid.

naturally occurring minerals: the monohydrate (whewellite), the dihydrate (weddellite) and the trihydrate (caoxite). The aluminium and iron (III) salts are highly insoluble in water, with the iron (II) salt being slightly more soluble. The solubility products are given in Table 2.29.

Table 2.27 Acid dissociation constants for oxalic acid.

Acid	Formula	Structure	Acid Dissociation Constant		Reference
			pK_{a2}	pK_{a1}	
Oxalic acid	HOOCCOOH		1.23 $(HOOCCOO)^-$	4.19 $(OOCCOO)^{2-}$	[51]
			Hydrogen oxalate	Oxalate	

Table 2.28 Stability constants of complexes formed between the oxalate ion relevant to cement chemistry.

Species	Reaction	Stability Constant, Log K	Reference
Ca(Oxalate)	$Ca^{2+} + C_2O_4^{2-} \rightleftharpoons Ca(C_2O_4)$	3.19	[52]
Ca(Oxalate)$_2^{2-}$	$Ca^{2+} + 2C_2O_4^{2-} \rightleftharpoons Ca(C_2O_4)_2^{2-}$	8.10	[52]
CaH(Oxalate)$^+$	$Ca^{2+} + C_2O_4^{2-} + H^+ \rightleftharpoons CaH(C_2O_4)^+$	6.03	[53]
CaH$_2$(Oxalate)$_2$	$Ca^{2+} + 2C_2O_4^{2-} + 2H^+ \rightleftharpoons CaH_2(C_2O_4)_2$	10.18	[53]
Al(Oxalate)$^+$	$Al^{3+} + C_2O_4^{2-} \rightleftharpoons Al(C_2O_4)^+$	7.7	[7]
Al(Oxalate)$_2^-$	$Al^{3+} + 2C_2O_4^{2-} \rightleftharpoons Al(C_2O_4)_2^-$	13.4	[7]
Al(Oxalate)$_3^{3-}$	$Al^{3+} + 3C_2O_4^{2-} \rightleftharpoons Al(C_2O_4)_3^{3-}$	17.0	[7]
AlH(Oxalate)$^{2+}$	$Al^{3+} + C_2O_4^{2-} + H^+ \rightleftharpoons AlH(C_2O_4)^{2+}$	7.5	[7]
AlOH(Oxalate)	$Al^{3+} + C_2O_4^{2-} + H_2O \rightleftharpoons AlOH(C_2O_4) + H^+$	2.6	[7]
AlOH(Oxalate)$_2^{2-}$	$Al^{3+} + 2C_2O_4^{2-} + H_2O \rightleftharpoons AlOH(C_2O_4)_2^{2-} + H^+$	6.8	[7]
Al(OH)$_2$(Oxalate)$^-$	$Al^{3+} + C_2O_4^{2-} + 2H_2O \rightleftharpoons Al(OH)_2(C_2O_4)^- + 2H^+$	−3.1	[7]
Fe(Oxalate)	$Fe^{2+} + C_2O_4^{2-} \rightleftharpoons Fe(C_2O_4)$	3.97	[7]
Fe(Oxalate)$_2^{2-}$	$Fe^{2+} + 2C_2O_4^{2-} \rightleftharpoons Fe(C_2O_4)^{2-}$	5.90	[7]
Fe(Oxalate)$^+$	$Fe^{3+} + C_2O_4^{2-} \rightleftharpoons Fe(C_2O_4)^+$	9.15	[7]
Fe(Oxalate)$_2^-$	$Fe^{3+} + 2C_2O_4^{2-} \rightleftharpoons Fe(C_2O_4)_2^-$	15.45	[7]
Fe(Oxalate)$_3^{3-}$	$Fe^{3+} + 3C_2O_4^{2-} \rightleftharpoons Fe(C_2O_4)_3^{3-}$	19.83	[7]
FeH(Oxalate)$^{2+}$	$Fe^{3+} + C_2O_4^{2-} + H^+ \rightleftharpoons FeH(C_2O_4)^{2+}$	4.35	[56]

Table 2.29 Solubility products of Ca, Al and Fe salts of oxalic acid.

Compound	Reaction	Solubility Product, $\log k_{sp}$	Reference	Molar Volume, cm^3/mol	Reference
Ca(Oxalate)	$Ca(C_2O_4) \rightleftharpoons Ca^{2+} + C_2O_4^{2-}$	−8.56	[54]	65.2	[57]
Ca(Oxalate).H$_2$O	$Ca(C_2O_4).H_2O \rightleftharpoons Ca^{2+} + C_2O_4^{2-} + H_2O$	−8.69	[54]	63.8	[58]
Ca(Oxalate).2H$_2$O	$Ca(C_2O_4).2H_2O \rightleftharpoons Ca^{2+} + C_2O_4^{2-} + 2H_2O$	−8.35	[54]	79.2	[58]
Ca(Oxalate).3H$_2$O	$Ca(C_2O_4).3H_2O \rightleftharpoons Ca^{2+} + C_2O_4^{2-} + 3H_2O$	−8.29	[54]	95.3	[59]
Al$_2$(Oxalate)$_3$.4H$_2$O	$Al_2(C_2O_4)_3.4H_2O \rightleftharpoons 2Al^{3+} + 3C_2O_4^{2-} + 4H_2O$	−33.46	[56]	Unknown	−
Fe(Oxalate).2H$_2$O	$Fe(C_2O_4).2H_2O \rightleftharpoons Fe^{2-} + C_2O_4^{2-} + 2H_2O$	−4.73	[55]	78.0	[60]
Fe$_2$(Oxalate)$_3$.5H$_2$O	$Fe_2(C_2O_4)_3.5H_2O \rightleftharpoons 2Fe^{3+} + 3C_2O_4^{2-} + 5H_2O$	−38.52	[56]	Unknown	−

The low solubility of calcium oxalate means that this solid phase persists even at low pH (Figure 2.27). Whilst the formation of complexes with oxalate ions destabilizes gibbsite and ferrihydrite, the low solubility of the Al and Fe oxalate salts means that solid phases, again, are present in significant quantities—at least at higher metal concentrations—across the full pH range (Figures 2.28 and 2.29).

Oxalic acid is one of the few organic compounds known to form complexes with silicon. However, the complex is weak, having a stability constant for the reaction $H_4SiO_4 + C_2O_4^{2-} \rightleftharpoons Si(C_6H_5O_7)(OH)_4^{2-}$ of 0.04 [61].

2.4.6 Pyruvic acid

Pyruvic acid is a relatively strong organic acid possessing three carbon atoms, one carboxylate group and one ketone (=O) making it the simplest of a group of acids known as the alpha-keto acids. Only the carboxylate group undergoes deprotonation (Table 2.30). Pyruvic acid forms a weak complex with calcium (Table 2.31), but no evidence exists of any complexes being formed between the pyruvate ion and either aluminium or iron. However, using a prediction technique based on the affinity of metal ions for the hydroxide ion ($\log K_1(OH^-)$) [62], it is unlikely that the log K forms of these stability constants would exceed a value of 3.0.

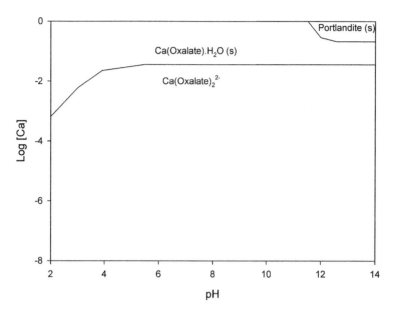

Figure 2.27 Solubility diagram for Ca in the presence of oxalic acid.

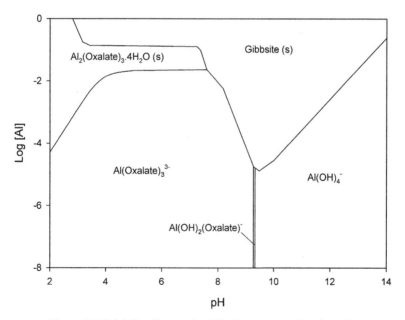

Figure 2.28 Solubility diagram for Al in the presence of oxalic acid.

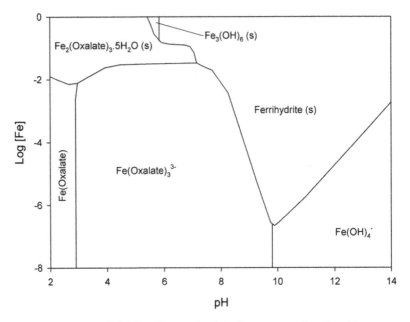

Figure 2.29 Solubility diagram for Fe in the presence of oxalic acid.

Table 2.30 Acid dissociation constants for pyruvic acid.

Acid	Formula	Structure	Acid Dissociation Constant	Reference
			pK$_a$	
Pyruvic	$CH_3C(=O)COOH$		2.50	[46]
			$CH_3C(=O)COO^-$	
			Pyruvate	

Table 2.31 Stability constants of complexes formed by calcium ions in water containing pyruvic acid.

Species	Reaction	Stability Constant, Log K	Reference
Ca(Pyruvate)	$Ca^{2+} + CH_3C(=O)COO^- \rightleftharpoons$ $Ca(CH_3C(=O)COO)^+$	1.08	[46]

Only the salt calcium pyruvate is known to form (Table 2.32), and quantitative data for this compound is seemingly lacking. However, on the assumption that calcium pyruvate is slightly soluble, a solubility of 5 g/l has been assumed, leading to the solubility diagram shown in Figure 2.30, indicating predominance of this compound at higher Ca concentrations over a relatively wide pH range.

2.4.7 Succinic acid

Succinic acid is structurally similar to oxalic acid in that it possesses two carboxylate groups at each end of the molecule. However, there is a longer chain of carbon atoms between these groups (Table 2.33). This has the effect of reducing the solubility of the acid relative to oxalic acid.

Succinic acid forms numerous complexes with aluminium, many of which involve additional deprotonation (Table 2.34). The succinate ion forms relatively insoluble salts with calcium (Table 2.35 and Figure 2.31). It also forms Fe(II) salts. Little is documented regarding the tetrahydrate, and the other salt, whilst of low solubility, appears to be formed only through hydrothermal synthesis techniques [68].

Table 2.32 Solubility products of Ca salts of pyruvic acid.

Compound	Reaction	Solubility Product, Log K_{sp}	Reference	Molar Volume, cm^3/mol	Reference
Ca(Pyruvate)	$Ca(CH_3C(=O)COO)_2.2.5H_2O \rightleftharpoons Ca^{2+} + 2CH_3C(=O)COO^- + 2.5H_2O$	'slightly soluble'	[63]	unknown	–

Table 2.33 Acid dissociation constants for succinic acid.

Acid	Formula	Structure	Acid Dissociation Constant		Reference
			pK_{a2}	pK_{a1}	
Succinic acid	$(CH_2)_2(COOH)_2$	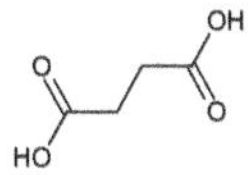	4.00 $HOOC(CH_2)_2COO^-$ Hydrogen succinate	5.42 $[(CH_2)_2(COO)_2]^{2-}$ Succinate	[53]

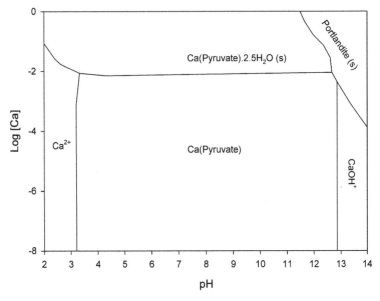

Figure 2.30 Solubility diagram for Ca in the presence of pyruvic acid.

2.4.8 Malic acid

Malic acid possesses two carboxylate groups and a hydroxyl group, with only the carboxylate groups normally undergoing deprotonation (Table 2.36). It forms an extremely varied series of complexes with aluminium and iron (III) (Table 2.37) where further deprotonation occurs. It forms iron (II) and calcium salts. Whilst there is little data on several of these, the dihydrate and trihydrate calcium salts are known to be of relatively low solubility (Table 2.38) leading to its predominance at higher calcium concentrations on the calcium solubility diagram (Figure 2.32).

2.4.9 Tartaric acid

Tartaric acid is a chiral compound, with the levotartaric (L-) form being the naturally occurring form. In addition, the mirror-image molecule, dextrotartaric acid (D-) and mesotartaric acid can be synthesized artificially. The molecule possesses two carboxylate groups, and two hydroxyl groups, although it is only the carboxylate groups which normally undergo deprotonation (Table 2.39).

Calcium forms complexes with both the partly and fully deprotonated molecule, whilst aluminium and iron only form complexes once the molecule is fully deprotonated. Calcium hydrogen tartrate and calcium

Table 2.34 Stability constants of complexes formed between the succinate ion relevant to cement chemistry.

Species	Reaction	Stability Constant, Log K	Reference
Ca			
Ca(Succinate)	$Ca^{2+} + [(CH_2)_2(COO)_2]^{2-} \rightleftharpoons$ $Ca[(CH_2)_2(COO)_2]$	1.20	[53]
CaH(Succinate)$^+$	$Ca^{2+} + [(CH_2)_2(COO)_2]^{2-} + H^+ \rightleftharpoons$ $Ca[HOOC(CH_2)_2COO]^+$	4.59	
Al			
Al(Succinate)$^+$	$Al^{3+} + [(CH_2)_2(COO)_2]^{2-} \rightleftharpoons$ $Al[(CH_2)_2(COO)_2]^+$	3.91	
AlH(Succinate)$^{2+}$	$Al^{3+} + [(CH_2)_2(COO)_2]^{2-} + H^+ \rightleftharpoons$ $Al[HOOC(CH_2)_2COO]^{2+}$	7.31	
Al$_2$(Succinate H$^+_{-3}$)$^+$	$2Al^{3+} + [(CH_2)_2(COO)_2]^{2-} \rightleftharpoons$ $Al_2[OOCCCHCOO]^+ + 3H^+$	–5.39	[64]
Al$_2$(Succinate H$^+_{-4}$)	$2Al^{3+} + [(CH_2)_2(COO)_2]^{2-} \rightleftharpoons$ $Al_2[OOC_4OO]$ $+ 4H^+$	–9.77	
Al$_3$(Succinate H$^+_{-1}$)$_2^{3+}$	$3Al^{3+} + 2[(CH_2)_2(COO)_2]^{2-} \rightleftharpoons$ $Al_3[OOCCHCH_2COO]_2^{3+} + 2H^+$	5.81	
Fe(II)			
Fe(Succinate)	$Fe^{2+} + [(CH_2)_2(COO)_2]^{2-} \rightleftharpoons$ $Fe[(CH_2)_2(COO)_2]$	2.4 (37°C)	[53]
Fe(III)			
Fe(Succinate)$^+$	$Fe^{3+} + [(CH_2)_2(COO)_2]^{2-} \rightleftharpoons$ $Fe[(CH_2)_2(COO)_2]^+$	6.88	[53]

tartrate are compounds of relatively low solubility, as is iron (II) tartrate (Table 2.40).

The solubility diagram for calcium in the presence of tartaric acid (Figure 2.33) shows calcium tartrate to persist as the dominant component of the system over a very wide pH range. In the case of iron, similar behaviour is observed, with solid ferrous tartrate persisting even under acidic conditions. Only qualitative solubility data was found for aluminium

Table 2.35 Solubility products of Ca and Fe salts of succinic acid.

Compound	Reaction	Solubility Product, $\log k_{sp}$	Reference	Molar Volume, cm³/mol	Reference
Calcium succinate trihydrate	$Ca[(CH_2)_2(COO)_2].3H_2O \rightleftharpoons$ $Ca^{2+} + [(CH_2)_2(COO)_2]^{2-} +$ $3H_2O$	−3.94	[74]	132.4	[65]
Calcium succinate monohydrate	$Ca[(CH_2)_2(COO)_2].H_2O \rightleftharpoons$ $Ca^{2+} + [(CH_2)_2(COO)_2]^{2-} +$ H_2O	−2.90	[74]	95.8	[66]
Iron (II) succinate tetrahydrate	$Fe[(CH_2)_2(COO)_2].4H_2O \rightleftharpoons$ $Fe^{2+} + [(CH_2)_2(COO)_2]^{2-} +$ H_2O	unknown	–	126.5	[67]
$Fe_5(OH)_2(C_4H_4O_4)_4$	$Fe_5(OH)_2((CH_2)_2(COO)_2)_4 \rightleftharpoons$ $5Fe^{2+} + 4[(CH_2)_2(COO)_2]^{2-}$ $+ 2OH^{2-}$	'insoluble'*	[68]	334.1	[68]

*probably only formed through hydrothermal synthesis.

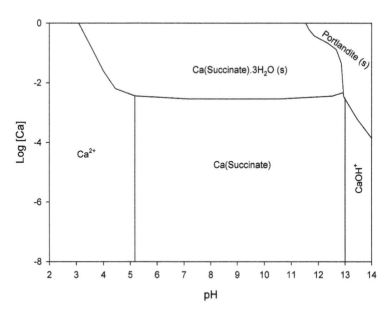

Figure 2.31 Solubility diagram for Ca in the presence of succinic acid.

Table 2.36 Acid dissociation constants for malic acid.

Acid	Formula	Structure	Acid Dissociation Constant		Reference
			pK_{a2}	pK_{a1}	
Malic acid	$CH_2CH(OH)$ $(COOH)_2$		3.24 $HOOCCH_2CH(OH)$ COO^- Hydrogen malate	4.68 $[CH_2CH(OH)$ $(COO)_2]^{2-}$ Malate	[33]

tartrate, but on the assumption that it is soluble in water, tartaric acid has the effect of solubilizing gibbsite under lower pH conditions (Figure 2.34).

In the case of iron, both the Fe(II) and Fe(III) salts are of low solubility (Table 2.41). However, it is the Fe(II) salt which has a significant effect on the solubility diagram (Figure 2.35)—it leads to a persistence of a solid iron phase to a low pH for higher iron concentrations.

2.4.10 Butyric acid

Butyric acid is a monocarboxylic acid, structurally similar to acetic and lactic acid. It has similar properties to these compounds (Table 2.42). It forms weak complexes with calcium and aluminium ions (Table 2.43) and soluble salts with calcium, aluminium, and iron (Table 2.44).

2.4.11 Fumaric acid

Fumaric acid is structurally similar to succinic acid, with the exception that the central carbon-carbon bond in its molecule is a double bond (Table 2.45). The lack of complexes identified as being formed by fumaric acid (Table 2.46) implies that this structural difference compromises the compound's ability to do so. It must be stressed that the absence of data partly indicates either a lack of experimental investigation into this compound or practical difficulties in conducting measurements. However, it is likely that the complexes formed are weak: using the same stability constant prediction method used for pyruvic acid, the estimated stability constant values are 2.3, 2.8 and 3.2 for 1:1 complexes of Al, Fe(II) and Fe(III).

Whilst the salts formed by fumaric acid are only sparingly to slightly soluble (Table 2.47), they do not appear as predominant species using the parameters used to form solubility plots in this chapter.

2.4.12 Gluconic acid

Gluconic acid possesses a single carboxylate group, but five hydroxyl groups. Normally only the carboxylate group takes part in deprotonation

Table 2.37 Stability constants of complexes formed between the malate ion relevant to cement chemistry.

Species	Reaction	Stability Constant, Log K	Reference
Ca			
Ca(Malate)	$Ca^{2+} + [C_4H_4O_5]^{2-} \rightleftharpoons Ca[C_4H_4O_5]$	1.96	
CaH(Malate)$^+$	$Ca^{2+} + [C_4H_4O_5]^{2-} + H^+ \rightleftharpoons Ca[C_4H_5O_5]^+$	5.77	[53]
Al			
Al(Malate)$^+$	$Al^{3+} + [C_4H_4O_5]^{2-} \rightleftharpoons Al[C_4H_4O_5]^+$	4.52	
AlH(Malate)$^{2+}$	$Al^{3+} + [C_4H_4O_5]^{2-} + H^+ \rightleftharpoons Al[C_4H_5O_5]^{2+}$	7.032	
AlH(Malate)$_2$	$Al^{3+} + 2[C_4H_4O_5]^{2-} + H^+ \rightleftharpoons Al[C_4H_5O_5]$ $[C_4H_4O_5]$	10.98	
Al(Malate H$^+_{-1}$)	$Al^{3+} + [C_4H_4O_5]^{2-} \rightleftharpoons Al[C_4H_3O_5] + H^+$	1.27	
Al$_2$(Malate H$^+_{-2}$)$^{2+}$	$2Al^{3+} + [C_4H_4O_5]^{2-} \rightleftharpoons Al_2[C_4H_2O_5]^{2+} + 2H^+$	0.56	
Al$_2$(Malate H$^+_{-3}$)$^+$	$2Al^{3+} + [C_4H_4O_5]^{2-} \rightleftharpoons Al_2[C_4H_1O_5]^+ + 3H^+$	–3.05	
Al$_2$(Malate H$^+_{-2}$) (Malate H$^+_{-1}$)$^-$	$2Al^{3+} + 2[C_4H_4O_5]^{2-} \rightleftharpoons Al_2[C_4H_2O_5][C_4H_3O_5]^-$ $+ 3H^+$	1.78	[69]
Al$_2$(Malate H$^+_{-2}$)$_2^{2-}$	$2Al^{3+} + 2[C_4H_4O_5]^{2-} \rightleftharpoons Al_2[C_4H_2O_5]_2^{2-} + 4H^+$	–4.46	
Al$_2$(Malate H$^+_{-1}$) (Malate)$_2^-$	$2Al^{3+} + 3[C_4H_4O_5]^{2-} \rightleftharpoons Al_2[C_4H_3O_5]$ $[C_4H_4O_5]_2^- + H^+$	12.79	
Al$_3$(Malate H$^+_{-1}$)$_4^{3-}$	$3Al^{3+} + 4[C_4H_4O_5]^{2-} \rightleftharpoons Al_3[C_4H_3O_5]_4^{3-} + 4H^+$	10.13	
Al$_4$(Malate H$^+_{-2}$) (Malate H$^+_{-1}$)$_3^-$	$4Al^{3+} + 4[C_4H_4O_5]^{2-} \rightleftharpoons Al_4[C_4H_2O_5]$ $[C_4H_3O_5]_3^- + 5H^+$	10.54	
Fe(II)			
Fe(Malate)	$Fe^{2+} + [C_4H_4O_5]^{2-} \rightleftharpoons Fe[C_4H_4O_5]$	2.6	[53]
Fe(III)			
Fe(Malate)$^+$	$Fe^{3+} + [C_4H_4O_5]^{2-} \rightleftharpoons Fe[C_4H_4O_5]^+$	7.13	
Fe$_2$(Malate H$^+_{-1}$)$_2$	$2Fe^{3+} + 2[C_4H_4O_5]^{2-} \rightleftharpoons Fe_2[C_4H_3O_5]_2 + 2H^+$	12.85	
Fe$_2$(Malate H$^+_{-1}$)$_2$ (Malate)$^{2-}$	$2Fe^{3+} + 3[C_4H_4O_5]^{2-} \rightleftharpoons$ $Fe_2[C_4H_3O_5]_2[C_4H_4O_5]^{2-} + 2H^+$	17.85	[70]
Fe$_3$(Malate)$_2$(Malate H$^+_{-1}$)$_2$(Malate H$^+_{-2}$)$^{5-}$	$3Fe^{3+} + 5[C_4H_4O_5]^{2-} \rightleftharpoons Fe_3[C_4H_4O_5]_2$ $[C_4H_3O_5]_2[C_4H_2O_5]^{5-} + 4H^+$	25.97	

Table 2.38 Solubility products of Ca and Fe salts of malic acid.

Compound	Reaction	Solubility Product, log k$_{sp}$	Reference	Molar Volume, cm^3/mol	Reference
Calcium malate trihydrate	$Ca(C_4H_4O_5).3H_2O \rightleftharpoons Ca^{2+} + (C_4H_4O_5)^{2-} + 3H_2O$	−3.49	[73]	128.0	[74]
Calcium malate dihydrate	$Ca(C_4H_4O_5).2H_2O \rightleftharpoons Ca^{2+} + (C_4H_4O_5)^{2-} + 2H_2O$	−2.56	[73]	115.0	[71]
Calcium hydrogen malate hexahydrate	$Ca(C_4H_5O_5)_2.6H_2O \rightleftharpoons Ca^{2+} + 2(C_4H_5O_5)^- + 6H_2O$	Unknown, assumed soluble		249.6	[72]
Iron (II) malate hydrate	$Fe(C_4H_4O_5).2.5H_2O \rightleftharpoons Fe^{2+} + (C_4H_4O_5)^{2-} + 2.5H_2O$	Unknown, assumed soluble		unknown	[73]
Iron (II) hydrogen malate tetrahydrate	$Fe(C_4H_5O_5)_2.4H_2O \rightleftharpoons Fe^{2+} + 2(C_4H_5O_5)^- + 4H_2O$	Unknown, assumed soluble		unknown	[73]

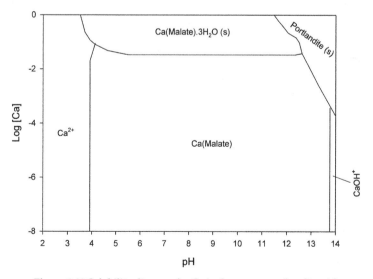

Figure 2.32 Solubility diagram for Ca in the presence of malic acid.

Table 2.39 Acid dissociation constants for tartaric acid.

Acid	Formula	Structure	Acid Dissociation Constant		Reference
			pK_{a2}	pK_{a1}	
Tartaric acid	HOOCCH(OH) CH(OH)COOH		3.06	4.37	[33]
		Hydrogen tartrate Tartrate			

Table 2.40 Stability constants of complexes formed by calcium, aluminium and iron(II) and (III) ions in water containing tartaric acid.

Complex	Reaction	Stability Constant, Log K	Reference
Ca			
	$Ca^{2+} + OOCCH(OH)CH(OH)CO_2^{2-} \rightleftharpoons$ $Ca(OOCCH(OH)CH(OH)CO_2)$	2.80	
	$Ca^{2+} + OOCCH(OH)CH(OH)CO_2^{2-} + H^+ \rightleftharpoons$ $Ca(HOOCCH(OH)CH(OH)CO_2)^+$	5.86	
Al	$Al^{3+} + 2OOCCH(OH)CH(OH)CO_2^{2-} \rightleftharpoons$ $Al(OOCC(OH)CH(OH)CO_2)_2^-$	9.37	[33]
Fe (II)	$Fe^{2+} + OOCCH(OH)CH(OH)CO_2^{2-} \rightleftharpoons$ $Fe(OOCCH(OH)CH(OH)CO_2)$	3.10	
Fe (III)	$Fe^{3+} + OOCCH(OH)CH(OH)CO_2^{2-} \rightleftharpoons$ $Fe(OOCCH(OH)CH(OH)CO_2)^+$	7.78	

(Table 2.48), but the hydroxyl groups may also deprotonate in the presence of iron(III) (Table 2.49). The salts of gluconic acid are soluble (Table 2.50). However, the calcium salt is close to the edge of this category and, whilst it does not appear on a solubility diagram using the parameters adopted in this chapter, could be precipitated at higher acid concentrations than the 0.1 mol/l used.

2.4.13 Propionic acid

Propionic acid has characteristics similar to most of the other monocarboxylic acids: it has a similar acid dissociation constant (Table 2.51) and forms relatively weak complexes (Table 2.52) and soluble salts (Table 2.53).

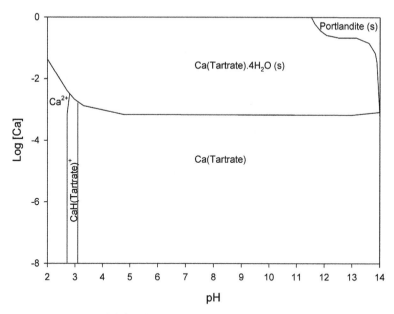

Figure 2.33 Solubility diagram for Ca in the presence of tartaric acid.

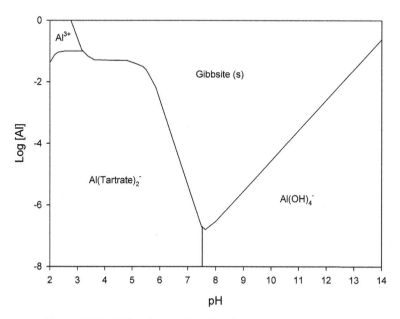

Figure 2.34 Solubility diagram for Al in the presence of tartaric acid.

Table 2.41 Solubility and molar volume data for compounds relevant to the interaction of hydrated Portland cement with tartaric acid.

Compound	Formula/Reaction	Solubility Product, log K_{sp}	Reference	Molar Volume, cm³/mol	Reference
Calcium tartrate tetrahydrate	Ca(OOCCH(OH)CH(OH) CO_2)·$4H_2O$ $\rightleftharpoons$ Ca^{2+} + OOCC(OH)CH(OH)CO_2^{2-} + $4H_2O$	−5.98 (D-tartaric)	[74]	141	[75]
Calcium hydrogen tartrate	Ca(HOOCCH(OH)CH(OH) $CO_2)_2$ $\rightleftharpoons$ Ca^{2+} + 2OOCC(OH) CH(OH)CO_2^{2-} + $2H^+$	−7.55	[76]	unknown	–
Aluminium tritartrate	Al$_2$(OOCCH(OH)CH(OH) $CO_2)_3$ $\rightleftharpoons$ 2Al^{3+} + 3OOCC(OH) CH(OH)CO_2^{2-}	'Soluble'	[77]	unknown	–
Iron (II) tartrate	Fe(OOCCH(OH)CH(OH) CO_2)·$2.5H_2O$ $\rightleftharpoons$ Fe^{2+} + OOCC(OH)CH(OH)CO_2^{2-} + $2.5H_2O$	−8.23	[78]	unknown	–
Iron (III) tartrate	Fe$_2$(OOCCH(OH)CH(OH) $CO_2)_3$·H_2O $\rightleftharpoons$ 2Fe^{3+} + 3OOCC(OH)CH(OH)CO_2^{2-} + H_2O	−14.525*		unknown	–

*based on solubility data provided by a number of chemical suppliers, but original source unclear.

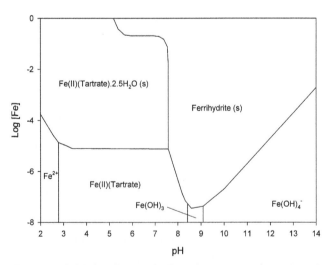

Figure 2.35 Solubility diagram for Fe in the presence of tartaric acid.

Table 2.42 Acid dissociation constants for butyric acid.

Acid	Formula	Acid Dissociation Constant pK_a	Reference
Butyric acid	$CH_3CH_2CH_2COOH$	4.82	[33]
	$CH_3CH_2CH_2CO_2^-$ Butyrate		

Table 2.43 Stability constants of complexes formed by calcium, aluminium and iron(II) and (III) ions in water containing butyric acid.

Complex	Reaction	Stability Constant, Log K	Reference
Ca			
	$Ca^{2+} + CH_3CH_2CH_2CO_2^- \rightleftharpoons$ $Ca(CH_3CH_2CH_2CO_2)^+$	0.94	[33]
Al			
	$Al^{3+} + CH_3CH_2CH_2CO_2^- \rightleftharpoons$ $Al(CH_2(OH)CO_2)^{2+}$	1.58	

Table 2.44 Solubility and molar volume data for compounds relevant to the interaction of hydrated Portland cement with butyric acid.

Compound	Formula/Reaction	Solubility Product, log K_{sp}	Reference	Molar Volume, cm³/mol	Reference
Calcium butyrate monohydrate	$Ca(CH_3CH_2CH_2CO_2)_2 \cdot H_2O$ $\rightleftharpoons Ca^{2+} + 2CH_3CH_2CH_2CO_2^-$ $+ H_2O$	0.04	[76]	179	[79]
Aluminium butyrate	$Al(CH_3CH_2CH_2CO_2)_3 \rightleftharpoons$ $Al^{3+} + 3CH_3CH_2CH_2CO_2^-$	Assumed soluble	–	Unknown	–
Iron (II) butyrate	$Fe(CH_3CH_2CH_2CO_2)_2 \rightleftharpoons$ $Fe^{2+} + 2CH_3CH_2CH_2CO_2^-$	Assumed soluble	–	Unknown	–
Iron (III) butyrate	$Fe(CH_3CH_2CH_2CO_2)_3 \rightleftharpoons$ $Fe^{3+} + 3CH_3CH_2CH_2CO_2^-$	Assumed soluble	–	Unknown	–

Table 2.45 Acid dissociation constants for fumaric acid.

Acid	Formula	Structure	Acid Dissociation Constant		Reference
			pK_{a2}	pK_{a1}	
Fumaric acid	$(CH)_2(COOH)_2$		3.053	4.494	[52]
			$HOOC(CH)_2COO^-$	$[(CH)_2(COO)_2]^{2-}$	
			Hydrogen fumarate	Fumarate	

Table 2.46 Stability constants of complexes formed by calcium ions in water containing fumaric acid.

Complex	Reaction	Stability Constant	Reference
Ca	$Ca^{2+} + C_4H_4O_4^{2-} \rightleftharpoons Ca(C_4H_4O_4)$	2.00	[53]

Table 2.47 Solubility and molar volume data for compounds relevant to the interaction of hydrated Portland cement with fumaric acid.

Compound	Formula/Reaction	Solubility Product, $\log k_{sp}$	Reference	Molar Volume, cm³/mol	Reference
Calcium fumarate	$Ca(C_4H_4O_4) \rightleftharpoons Ca^{2+} + C_4H_4O_4^{2-}$	−1.77	[80]	unknown	–
Calcium fumarate trihydrate	$Ca(C_4H_4O_4)\cdot3H_2O \rightleftharpoons Ca^{2+} + C_4H_4O_4^{2-} + 3H_2O$	−1.99	[80]	122	[81]
Iron (II) fumarate	$Fe(C_4H_4O_4) \rightleftharpoons Fe^{2+} + C_4H_4O_4^{2-}$	−4.17	[80]	70	[80]

Table 2.48 Acid dissociation constants for gluconic acid.

Acid	Formula	Acid Dissociation Constant	Reference
		pK_a	
Gluconic acid	$HOCH_2(CHOH)_4COOH$	3.56	[33]
		$HOCH_2(CHOH)_4COO^-$	
		gluconate	

Table 2.49 Stability constants of complexes formed by calcium, aluminium and iron(II) and (III) ions in water containing gluconic acid.

Complex	Reaction	Stability Constant, Log K	Reference
Ca	$Ca^{2+} + C_6H_{11}O_7^- \rightleftharpoons Ca(C_6H_{11}O_7)^+$	1.21	[33]
Al	$Al^{3+} + C_6H_{11}O_7^- \rightleftharpoons Al(C_6H_{11}O_7)^{2+}$	2.4	[82]
Fe (II)	$Fe^{2+} + C_6H_{11}O_7^- \rightleftharpoons Fe(C_6H_{11}O_7)^+$	1.0	
Fe (III)	$Fe^{3+} + C_6H_{11}O_7^- \rightleftharpoons Fe(C_6H_{11}O_7)^{2+}$	−3.1	[33, 83]
	$Fe^{3+} + C_6H_{11}O_7^- \rightleftharpoons Fe(C_6H_{10}O_7)^+ + H^+$	−0.8	
	$Fe^{3+} + C_6H_{11}O_7^- \rightleftharpoons Fe(C_6H_9O_7) + 2H^+$	1.5	
	$Fe^{3+} + C_6H_{11}O_7^- \rightleftharpoons Fe(C_6H_8O_7)^- + 3H^+$	5.5	
	$Fe^{3+} + C_6H_{11}O_7^- \rightleftharpoons Fe(C_6H_7O_7)^{2-} + 4H^+$	18.8	

Table 2.50 Solubility and molar volume data for compounds relevant to the interaction of hydrated Portland cement with gluconic acid.

Compound	Formula/Reaction	Solubility Product, log K_{sp}	Reference	Molar Volume, cm³/mol	Reference
Calcium gluconate monohydrate	$Ca(C_6H_{11}O_7)_2 \cdot H_2O \rightleftharpoons Ca^{2+} + 2C_6H_{11}O_7^- + H_2O$	−2.44	[84]	unknown	
Aluminium gluconate hydroxide	$Al(OH)(C_6H_{11}O_7)_2 \rightleftharpoons Al^{3+} + 3 2C_6H_{11}O_7^- + OH^-$	Assumed soluble	[85]	unknown	
Iron (II) gluconate dihydrate	$Fe(C_6H_{11}O_7)_2 \cdot 2H_2O \rightleftharpoons Fe^{2+} + 2C_6H_{11}O_7^-$	'Soluble'	[8]	unknown	

Table 2.51 Acid dissociation constants for propionic acid.

Acid	Formula	Acid Dissociation Constant pK$_a$	Reference
Propionic acid	CH_3CH_2COOH	4.874	[8]

$CH_3CH_2CO_2^-$
propionate

Table 2.52 Stability constants of complexes formed by calcium, aluminium and iron(III) ions in water containing propionic acid.

Complex	Reaction	Stability Constant, Log K	Reference
Ca			
	$Ca^{2+} + CH_3CH_2CO_2^- \rightleftharpoons Ca(CH_3CH_2CO_2)^+$	0.93	[33]
Al			
	$2Al^{3+} + CH_3CH_2COOH + 2H_2O \rightleftharpoons$ $Al_2(OH)_2(CH_3CH_2CO_2)^{3+} + 3H^+$	−8.04	[86]
Fe(III)			
	$Fe^{3+} + CH_3CH_2CO_2^- \rightleftharpoons Fe(CH_3CH_2CO_2)^{2+}$	4.01	[33]

Table 2.53 Solubility and molar volume data for compounds relevant to the interaction of hydrated Portland cement with propionic acid.

Compound	Formula/Reaction	Solubility Product, log k_{sp}	Reference	Molar Volume, cm³/mol	Reference
Calcium propionate	$Ca(CH_3CH_2CO_2)_2 \rightleftharpoons Ca^{2+} +$ $2CH_3CH_2CO_2^{2-}$	1.91	[87]	Unknown	–
Iron (III) propionate	$Fe_3O(CH_3CH_2CO_2)_7(CH_3COOH) \rightleftharpoons Fe^{3+} + 8CH_3CO_2^- + H_2O$	'soluble'	[88]	Unknown	–

2.4.14 Citric acid

Citric acid is a carboxylic acid possessing three carboxylate groups and an additional –OH group. Only the carboxylate groups normally undergo deprotonation (Table 2.54). The acid forms strong complexes with Ca, Al and Fe ions (Table 2.55).

It is capable of forming a number of salts with calcium, the three most likely to be encountered shown in Table 2.56. Of these salts, the least soluble is calcium citrate tetrahydrate ($Ca_3(Citrate)_2·4H_2O$), which is observed as occupying the higher calcium concentration range of the calcium solubility diagram in Figure 2.36. Aluminium and iron (III) salts are also reported in the literature, although they are soluble.

The strength of the complexes formed between iron and aluminium are such that they limit the normal dominance of gibbsite and ferrihydrite (Figures 2.37 and 2.38).

Table 2.54 Acid dissociation constants for citric acid.

Acid	Formula	Structure	Acid Dissociation Constant			Reference
			pK_{a3}	pK_{a2}	pK_{a1}	
Citric acid	$C_6H_8O_7$		3.13	4.76	6.40	[33]
			Dihydrogen citrate	Hydrogen citrate	Citrate	

Table 2.55 Stability constants of complexes formed by calcium, aluminium and iron(II) and (III) ions in water containing citric acid.

Complex	Reaction	Stability Constant, Log K	Reference
Ca			
	$Ca^{2+} + C_6H_5O_7^{3-} \rightleftharpoons Ca(C_6H_5O_7)^-$	4.87	
	$Ca^{2+} + C_6H_5O_7^{3-} + H^+ \rightleftharpoons Ca(C_6H_6O_7)$	9.26	
	$Ca^{2+} + C_6H_5O_7^{3-} + 2H^+ \rightleftharpoons CaH(C_6H_7O_7)$	12.26	[33]
Al	$Al^{3+} + C_6H_5O_7^{3-} \rightleftharpoons Al(C_6H_5O_7)$	9.97	
	$Al^{3+} + 2C_6H_5O_7^{3-} \rightleftharpoons Al(C_6H_5O_7)_2^{3-}$	14.80	
	$Al^{3+} + C_6H_5O_7^{3-} + H^+ \rightleftharpoons Al(C_6H_6O_7)^+$	12.85	
Fe (II)	$Fe^{2+} + C_6H_5O_7^{3-} \rightleftharpoons Fe(C_6H_5O_7)^-$	6.1	
	$Fe^{2+} + C_6H_5O_7^{3-} + H^+ \rightleftharpoons Fe(C_6H_6O_7)$	10.2	
Fe (III)	$Fe^{3+} + C_6H_5O_7^{3-} \rightleftharpoons Fe(C_6H_5O_7^{3-})$	13.1	
	$Fe^{3+} + C_6H_5O_7^{3-} + H^+ \rightleftharpoons Fe(C_6H_6O_7)^+$	14.4	

Citric acid is known to form complexes with silicon. However, the strength of the complex is weak: the stability constant for the reaction $H_4SiO_4 + C_6H_5O_7^{3-} \rightleftharpoons Si(C_6H_5O_7)(OH)_4^{3-}$ is 0.11 [61].

2.5 Deterioration Mechanisms

The previous section has outlined the manner in which various acids interact chemically with the chemical constituents of conventional cements.

Table 2.56 Solubility and molar volume data for compounds relevant to the interaction of hydrated Portland cement with citric acid.

Compound	Formula/Reaction	Solubility Product, $\log k_{sp}$	Reference	Molar Volume, cm^3/mol	Reference
Calcium citrate tetrahydrate	$Ca_3(C_6H_5O_7)_2 \cdot 4H_2O \rightleftharpoons 3Ca^{2+} + 2C_6H_5O_7^{3-} + 4H_2O$	−14.94*	[74]	285*	[89]
Calcium hydrogen citrate tetrahydrate	$Ca(C_6H_6O_7) \cdot 4H_2O \rightleftharpoons Ca^{2+} + C_6H_5O_7^{3-} + 4H_2O + H^+$	−0.33	[76]	157	[90]
Calcium dihydrogen citrate trihydrate	$Ca(C_6H_6O_7)_2 \cdot 3H_2O \rightleftharpoons Ca^{2+} + 2C_6H_5O_7^{3-} + 3H_2O + 4H^+$	0.77	[76]	unknown	–
Aluminium citrate	$Al(C_6H_5O_7) \rightleftharpoons Al^{3+} + C_6H_5O_7^{3-}$	1.26	[91]	unknown	–
Iron (III) citrate pentahydrate	$Fe(C_6H_5O_7) \cdot 5H_2O \rightleftharpoons Fe^{3+} + C_6H_5O_7^{3-} + 5H_2O$	'slightly soluble'	[74]	unknown	–

*These values are for the Earlandite form of calcium citrate tetrahydrate. At least one other form appears to exist. This other form is seemingly the one formed in contact with cement [92].

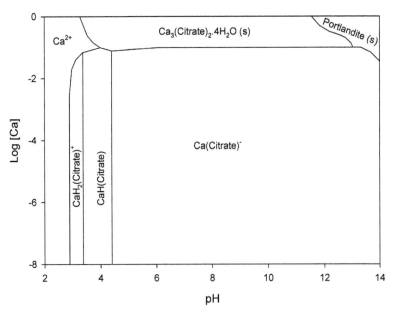

Figure 2.36 Solubility diagram for Ca in the presence of citric acid.

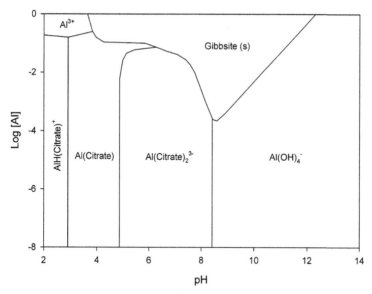

Figure 2.37 Solubility diagram for Al in the presence of citric acid.

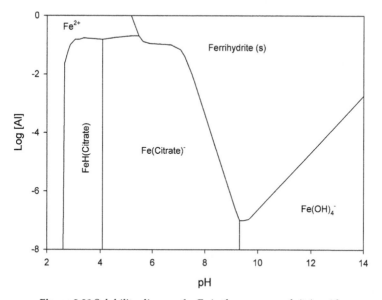

Figure 2.38 Solubility diagram for Fe in the presence of citric acid.

However, understanding the manner in which these interactions impact on the integrity of cement and concrete make it necessary to examine the mechanisms which are effective when acid comes into contact with such materials. So far, the interactions we have examined have isolated certain cement components, but we must now consider all the constituents together. Moreover, interaction has exclusively examined interaction in chemical terms, but there are also physical aspects to deterioration.

2.5.1 Deterioration from leaching

Concrete exposed to acids may undergo deterioration as a result of the accelerated leaching of chemical constituents in the cement matrix, and possibly also in the aggregate. This can occur by two mechanisms: *acidolysis* and *complexolysis*.

Acidolysis involves the reaction of hydration products with the acid to yield soluble ions. Thus, any of the acids discussed in the previous section which form soluble salts will cause deterioration by acidolysis. Of most significance is the leaching of portlandite:

$$Ca(OH)_2 + 2H^+ \rightarrow Ca^{2+} + 2H_2O.$$

This is because—as has been seen from the solubility diagrams—portlandite undergoes this reaction at relatively high pH. Where acidolysis is the main mechanism of deterioration this will normally lead to the complete dissolution of this hydration product. This has profound implications for the cement matrix, because portlandite tends to be present as relatively large crystals, leaving pores which weaken the material considerably [93].

In contrast—again evident from the solubility diagrams—the acid decomposition of the AFm and AFt phases leads to the formation of amorphous gibbsite and ferrihydrite, which normally persist to a low pH.

CSH gel undergoes a process of decalcification: calcium is progressively removed from the gel by the process of acidolysis, leading to a drop in the Ca/Si ratio until only silica gel remains. This also leads to a decrease in strength, but is generally less than that from the loss of portlandite. For this reason, it is often observed that cements containing a lower portlandite content (such as those containing proportions of other cementitious materials) perform better than Portland cement alone where acidolysis is the main mechanism of deterioration.

Calcium aluminate cements undergo a different process, with the precipitation of amorphous gibbsite playing a major role. If it is assumed that the cement is fully converted, then the cement will initially take part in the following reaction:

$$Ca_3Al_2O_9.6H_2O + 6H^+ \rightarrow 3Ca^{2+} + 2Al(OH)_3 + 6H_2O$$

The gibbsite formed will eventually undergo acidolysis at much lower pH:

$$2Al(OH)_3 + 6H^+ \rightarrow 2Al^{3+} + 6H_2O$$

Complexolysis is an enhanced leaching of the cement resulting from the formation of strong complexes. Complex formation allows more of the solid phases to dissolve, although it should be stressed that acidolysis must occur prior to complex formation, and so the two mechanisms occur in tandem. A good example of complexolysis is shown in Figure 2.37 where the formation of strong complexes between citric acid and aluminium ions leads to a destabilization of gibbsite.

For both mechanisms, the process of deterioration will also be controlled by the movement of acid into the concrete itself. This will occur by diffusion and the rate will be dependent on the porosity in the concrete and the extent to which porosity is increased by the deterioration process.

The simultaneous role of both chemical reactions and mass transport makes the acid attack process a potentially complex one. For this reason, where deterioration by acids is discussed in subsequent chapters, the use of a technique known as geochemical modelling will be employed to illustrate the outcomes of attack from different acids on different cement types. This has been done using the computer program PHREEQC [94], which uses stability constant and solubility product data along with mass transport calculations to simulate the processes occurring when solid minerals (i.e., the cement hydration products) and aqueous solutions come into contact.

2.5.2 Deterioration from expansive reaction products

Another way in which acids can cause damage to concrete is through the precipitation of expansive solid salts. As a general rule, a salt will be expansive if it occupies a larger volume than the substance it replaces through chemical reaction. This can be deduced by examining the molar volume of the reactant and product. These values, where available, have been included in the salt formation tables provided throughout Section 2.4.

Thus, when gypsum is formed as a result of the reaction of sulfate ions with portlandite, the reaction takes the form:

$$Ca(OH)_2 + SO_4^{2-} + H_2O \rightarrow CaSO_4.2H_2O + 2e^-$$

one mole of portlandite (with a molar volume of 32.9 cm^3/mol) will be converted into one mole of gypsum (molar volume = 74.5 cm^3/mol). The resulting increase in volume, when occurring in the confines of pores and cracks in concrete may be enough to cause further cracking and fragmentation of the cement matrix.

Whilst this general rule of thumb is useful in identifying potential issues with specific acids, the process of expansive salt formation is complex.

Various factors will play a role, including the solubility of the salt and the pH range over which it remains stable. As a result, salt formation requires individual investigation for specific acids. Where data are available, this is attempted in Chapter 4, which looks at fungal deterioration of concrete. This is because fungi are largely responsible for the production of acids involved for this type of deterioration.

2.6 References

[1] Drever JI (1997) The Geochemistry of Natural Waters. 3rd Ed., Prentice Hall, Upper Saddle River.

[2] Davies CW (1962) Ion Association. Butterworths, Washington.

[3] Truesdell AH and Jones BF (1973) WATEQ, a computer program for calculating chemical equilibria of natural waters. US Geological Survey Journal of Research **2**: 233–248.

[4] Chen JJ, Thomas J, Taylor HFW and Jennings HM (2004) Solubility and structure of calcium silicate hydrate. Cement and Concrete Research **34(9)**: 1499–1519.

[5] Winnefeld F and Lothenbach B (2010) Hydration of calcium sulfoaluminate cements—experimental findings and thermodynamic modelling. Cement and Concrete Research **40(8)**: 1239–1247.

[6] Glasser FP (1992) Chemistry of cement-solidified waste forms. pp. 1–39. *In*: Spence RD (ed.). Chemistry and Microstructure of Solidified Waste Forms. Lewis Publishers, Boca Raton, FL, USA.

[7] U.S. Environmental Protection Agency (1999) MINTEQA2/PRODEFA2, A Geochemical Assessment Model for Environmental Systems—User Manual Supplement for Version 4.0. National Exposure Research Laboratory, Athens, GA, USA.

[8] Haynes WM (2014) CRC Handbook of Chemistry and Physics. 95th Ed., CRC Press, Boca Raton, Florida, USA.

[9] Marshall WL and Jones EV (1966) Second dissociation constant of sulfuric acid from 25 to 350° evaluated from solubilities of calcium sulfate in sulfuric acid solutions. Journal of Physical Chemistry **70**: 4028–4040.

[10] Damidot D and Glasser FP (1993) Thermodynamic investigation of the $CaO-Al_2O_3-CaSO_4-H_2O$ system at 25°C and the influence of Na_2O. Cement and Concrete Research **23**: 221–238.

[11] Balonis M and Glasser FP (2009) The density of cement phases. Cement and Concrete Research **39**: 733–739.

[12] Roberts WL (1990) Encyclopedia of Minerals. 2nd Ed., Van Nostrand Reinhold.

[13] Möschner G, Lothenbach B, Rose J, Ulrich A, Figi R and Kretzschmar R (2008) Solubility of Fe–ettringite ($Ca_6[Fe(OH)_6]2(SO_4)_3.26H_2O$). Geochimica et Cosmochimica Acta **72**: 1–18.

[14] Dilnesa BZ, Lothenbach B, Renaudin G, Wichser A and Kulik D (2014) Synthesis and characterization of hydrogarnet $Ca_3(Al_xFe_{1-x})2(SiO_4)_y(OH)_{4(3-y)}$. Cement and Concrete Research **59**: 96–111.

[15] Singh SS (1980) Thermodynamic properties of synthetic basic aluminite $[Al_4(OH)_{10}SO_4.5H_2O]$ from solubility data. Canadian Journal of Soil Science **60**: 381–384.

[16] van Breemen N (1973) Dissolved aluminum in acid sulfate soils and in acid mine waters. Soil Science Society of America Proceedings **37**: 694–697.

[17] Jones AM, Collins RN and Waite TD (2011) Mineral species control of aluminum solubility in sulfate-rich acidic waters. Geochimica et Cosmochimica Acta **75**: 965–977.

[18] Kolthoff I (1959) Treatise on Analytical Chemistry. Interscience Encyclopedia, New York.

[19] Bailar JC and Trotman-Dickenson AF (1973) Comprehensive Inorganic Chemistry. Pergamon Press, Oxford.

[20] Balonis M (2010) The influence of inorganic chemical accelerators and corrosion inhibitors on the mineralogy of hydrated portland cement systems. Ph.D. thesis, University of Aberdeen.

[21] Renaudin G and Francois M (1999) The lamellar double-hydroxide (LDH) compound with composition $3CaO.Al_2O_3.Ca(NO_3)_2.10H_2O$. Acta Crystallographica C **55:** 835–838.

[22] Speight J (2005) Lange's Handbook of Chemistry. 16th Ed., McGraw-Hill, London.

[23] Ribar B, Divjakovic V, Herak R and Prelesnik B (1973) A new crystal structure study of $Ca(NO_3)_2.4H_2O$. Acta Crystallographica B **29:** 1546–1548.

[24] Herpin P and Sudarsanen K (1965) Crystal structure of aluminum nitrate hydrate. Bulletin de la Societe Francaise de Mineralogie et de Cristallographie **88:** 595–601.

[25] Schmidt H, Asztalos A, Bok F and Voigt W (2012) New iron(III) nitrate hydrates: $Fe(NO_3)_3.xH_2O$ with x = 4, 5 and 6. Acta Crystallographica C **68:** i29–i33.

[26] Park J-Y and Lee Y-N (1988) Solubility and decomposition kinetics of nitrous acid in aqueous solution. Journal of Physical Chemistry **92:** 6294–6302.

[27] Brooker MH and DeYoung BS (1980) Measurement of the stability constants for aqueous calcium and magnesium nitrite by Raman spectroscopic methods. Journal of Solution Chemistry **9:** 279–288.

[28] Plummer LN and Busenberg E (1982) The solubilities of calcite, aragonite and vaterite in CO_2-H_2O solutions between 0 and 90°C, and an evaluation of the aqueous model of the system $CaCO_3$-CO_2-H_2O. Geochimica and Cosmochimica Acta **46:** 1011–1040.

[29] Rodriguez-Blanco JD, Shaw S and Benning LG (2011) The kinetics and mechanisms of amorphous calcium carbonate (ACC) crystallization to calcite, via vaterite. Nanoscale **3:** 265–271.

[30] Graf DL (1961) Crystallographic tables for the rhombohedral carbonates. American Mineralogist **46:** 1283–1316.

[31] Reardon EJ (1990) An ion interaction-model for the determination of chemical-equilibria in cement water-systems. Cement and Concrete Research **20:** 175–192.

[32] Lothenbach B, Matschei T, Möschner G and Glasser FP (2008) Thermodynamic modelling of the effect of temperature on the hydration and porosity of Portland cement. Cement and Concrete Research **38:** 1–18.

[33] Martell AE and Smith RM (2001) Critical Selected Stability Constants of Metal Complexes Database, Version 6.0 for Windows; National Institute of Standards and Technology, April.

[34] International Union of Pure and Applied Chemistry, IUPAC-NIST Solubility Database, Version 1.1, National Institute of Science and Technology, 2015. http://srdata.nist.gov/solubility/index.aspx.

[35] Nitta I and Osaki K (1948) Crystal structure of calcium formate. X-sen Osaka University **5:** 37–42.

[36] Tian Y-Q, Zhao Y-M, Xu H-J and Chi C-Y (2007) CO_2 template synthesis of metal formates with a ReO_3 net. Inorganic Chemistry **46:** 1612–1616.

[37] Weber G (1980) Iron(II) formate dihydrate. Acta Crystallographica B **36:** 3107–3109.

[38] Pöllmann H, Kaden R and Stöber S (2014) Crystal structures and XRD data of new calcium aluminate cement hydrates. pp. 75–85. *In*: Fentiman CH, Mangabhai RJ and Scrivener KL (eds.). Calcium Aluminates: Proceedings of the International Conference 2014. IHS BRE Press.

[39] Saury C, Boistelle R, Dalemat F and Bruggeman J (1993) Solubilities of calcium acetates in the temperature range 0–100°C. Journal of Chemical Engineering Data **38:** 56–59.

[40] Klop EA, Schouten A, van der Sluis P and Spek AL (1984) Structure of calcium acetate monohydrate, $Ca(C_2H_3O_2)_2.H_2O$. Acta Crystallographica C **40:** 51–53.

[41] Balarew Chr, Stoilova D and Demirev L (1974) Untersuchung einiger dreistoffsysteme vom typ $Me(OCOCH_3)_2$-CH_3COOH-H_2O bei 25°C (Me = Ni, Co, Mg, Mn, Ca). Zeitschrift für Anorganische und Allgemeine Chemie **410:** 75–87.

[42] Klop EA and Spek AL (1984) Structure of calcium hydrogen triacetate monohydrate, $CaH(C_2H_3O_2)_3.H_2O$. Acta Crystallographica C **40:** 1817–1819.

[43] Lewis RJ (2007) Hawley's Condensed Chemical Dictionary. 15th Ed., Wiley-Interscience, Hoboken NJ, USA.

[44] Alcala R and Garcia JF (1973) Revista de la Academia de Ciencias Exactas. Fisico-Quimicas y Naturales de Zaragoza **28**: 303.

[45] Hamada YZ, Carlson B and Dangberg J (2005) Interaction of malate and lactate with chromium(III) and iron(III) in aqueous solutions. Synthesis and Reactivity in Inorganic, Metal-Organic and Nano-Metal Chemistry **35**: 515–522.

[46] Dawson RMC (1959) Data for Biochemical Research. Clarendon Press, Oxford.

[47] Marklund E, Sjöberg S, Öhman L-O, Salvatore F, Niinistö L and Volden HV (1986) Equilibrium and structural studies of silicon(IV) and aluminium(III) in aqueous solution. 14. Speciation and equilibria in the aluminium(III)lactic acid-OH-system. Acta Chemica Scandinavica **40**: 367–373.

[48] Apelblat A, Manzurola E, van Krieken J and Nanninga GL (2005) Solubilities and vapour pressures of water over saturated solutions of magnesium-l-lactate, calcium-l-lactate, zinc-l-lactate, ferrous-l-lactate and aluminum-l-lactate. Fluid Phase Equilibria **236**: 162–168.

[49] Bombi GG, Corain B, Sheikh-Osman AA and Giovanni C Valle (1990) The speciation of aluminum in aqueous solutions of aluminum carboxylates. Part I. X-ray molecular structure of $Al[OC(O)CH(OH)CH_3]_3$. Inorganica Chimica Acta **171**: 79–83.

[50] Adu-Wusu K (2012) Literature review on impact of glycolate on the 2H evaporator and the effluent treatment facility (ETF). Savannah River National Laboratory, Aiken SC, USA.

[51] Sillen LG and Martell AE (1971) Stability Constants of Metal-Ion Complexes, Supplement No. 1. The Chemical Society, London.

[52] Martell AE and Smith RM (1989) Critical Stability Constants, Vol. 6 Second Supplement. Plenum, London.

[53] Martell AE and Smith RM (1977) Critical Stability Constants, Vol. 3 Other Organic Ligands. Plenum, London.

[54] Streit J, Tran-Ho LC and Koenigsberger E (1998) Solubility of calcium oxalate hydrates in sodium chloride solutions and urine-like liquors. Monatshefte für Chemie **129**: 1225–1236.

[55] Weast RC, Astel MJ and Beyer WH (1986) CRC Handbook of Chemistry and Physics. 67th Ed., CRC Press, Boca Raton, Florida, USA.

[56] Christodoulou E, Panias D and Paspaliaris I (2001) Calculated solubility of trivalent iron and aluminum in oxalic acid solutions at 25°C. Canadian Metallurgical Quarterly **40**: 421–432.

[57] Hochrein O, Thomas A and Kniep R (2008) Revealing the structure of anhydrous calcium oxalate, $Ca[C_2O_4]$, by a combination of atomistic simulation and Rietveld refienement. Zeitschrift Anorganische Allgemeine Chemie **634**: 1826–1829.

[58] Tazzoli V and Domeneghetti C (1980) The crystal structures of whewellite and wheddelite: re-examination and comparison. American Mineralogist **65**: 327–334.

[59] Deganello S, Kampf AR and Moore PB (1981) The crystal structure of calcium oxalate trihydrate: $Ca(H_2O)_3(C_2O_4)$. American Mineralogist **66**: 859–865.

[60] Echigo T and Kimata M (2008) Single crystal X-ray diffraction and spectroscopic studies on humboldtine and lindbergite: weak Jahn-Teller effect of Fe^{2+} ion. Physical Chemistry of Minerals **35**: 467–475.

[61] Öhman LO, Nordin A, Sedeh IF and Sjöberg S (1991) Equilibrium and structural studies of silicon(IV) and aluminium(III) in aqueous solution. 28. Formation of soluble silicic acid–ligand complexes as studied by potentiometric and solubility measurements. Acta Chemica Scandinavica **45**: 335–341.

[62] Hancock RD and Martell AE (1989) Ligand design for selective complexation of metal ions in aqueous solution. Chemical Reviews **89(8)**: 1875–1914.

[63] Panel on Food Additives and Nutrient Sources, Calcium acetate, calcium pyruvate, calcium succinate, magnesium pyruvate magnesium succinate and potassium malate

added for nutritional purposes to food supplements. The EFSA Journal **1088**: 2009, pp. 1–25.

[64] Alliey N, Venturini-Soriano M and Berthon G (1996) Aluminum-succinate complex equilibria and their potential implications for aluminum metabolism. Annals of Clinical and Laboratory Science **26**: 122–132.

[65] Karipides A and Reed AT (1980) The structure of diaquasuccinatocalcium(II) monohydrate. Acta Crystallographica B **36**: 1377–1381.

[66] Mathew M, Takagi S, Fowler BO and Markovic M (1994) The crystal structure of calcium succinate monohydrate. Journal of Chemical Crystallography **24**: 437–440.

[67] Xu T-G, Xu D-J, Wu J-Y and Chiang MY (2002) *catena*-Poly[[tetraaquairon(II)]-[mu]-succinato-[kappa]2O:O']. Acta Crystallographica C **58**: m615–m616.

[68] Kim Y-J and Jung D-Y (1999) $Fe_5(OH)_2(C_4H_4O_4)_4$: hydrothermal synthesis of microporous iron(II) dicarboxylate with an inorganic framework. Bulletin Korean Chemical Society **20**: 830.

[69] Berthon G (2002) Aluminium speciation in relation to aluminium bioavailability, metabolism and toxicity. Coordination Chemistry Reviews **228(2)**: 319–341.

[70] Timberlake CF (1964) Iron–malate and iron–citrate complexes. Journal of the Chemical Society 5078–5085.

[71] Bränden C-I and Söderberg B-O (1966) The crystal structure of Ca-malate-dihydrate, $CaC_4H_4O_5.2H_2O$. Acta Chemica Scandinavica **20**: 730–738.

[72] Lenstra TH and Van Havere W (1980) Calcium di(hydrogen 1-malate) hexahydrate. Acta Crystallographica **B36**: 156–158.

[73] Schmittler H (1968) Cell dimensions of some salts of malic acid. Acta Crystallographica **B24**: 983–984.

[74] Bertron A and Duchesne J (2013) Attack of cementitious materials by organic acids in agricultural and agrofood effluents. pp. 131–173. *In*: Alexander M, De Belie N and Bertron A (eds.). Performance of Cement—based Materials in Aggressive Aqueous Environments. RILEM State-of-the-Art Report TC 211-PAE. Springer, Dordrecht, Netherlands.

[75] Hawthorn FC, Borys I and Ferguson RB (1982) Structure of calcium tartrate tetrahydrate. Acta Crystallographica B **38**: 2461–2463.

[76] Seidell A (1919) Solubilities of Inorganic and Organic Compounds. 2nd Ed., Van Nostrand, New York.

[77] O'Neil MJ (2001) The Merck Index: An Encyclopedia of Chemicals, Drugs, and Biologicals. 13th Ed., Merck, Whitehouse Station, NJ, USA.

[78] Cayot P, Guzun-Cojocaru T and Cayot N (2013) Iron fortification of milk and dairy products. pp. 75–89. *In*: Preedy VR, Srirajaskanthan R and Patel VB (eds.). Handbook of Food Fortification and Health: From Concepts to Public Health Applications. Volume 1 Springer, New York.

[79] Valora A, Reguera E and Sanchez-Sinencio F (2002) Synthesis and X-ray diffraction study of calcium salts of some carboxylic acids. Powder Diffraction **17**: 13–18.

[80] Weast R (1966) CRC Handbook of Chemistry and Physics. 47th Ed., Chemical Rubber Co, Cleveland, OH, USA.

[81] Gupta MP, Prasad SM, Sahu RG and Sahu BN (1972) The crystal structure of calcium fumarate trihydrate $CaC_4H_2O_4.3H_2O$. Acta Crystallographica **B28**: 135–139.

[82] Pallagi A, Tasi ÁG, Peintler G, Forgo P, Pálinkó I and Sipos P (2013). Complexation of Al(III) with gluconate in alkaline to hyperalkaline solutions: formation, stability and structure. Dalton Transactions **42(37)**: 13470–13476.

[83] Pecsok RL and Sandera J (1955) The gluconate complexes. II. The ferric-gluconate system. Journal of the American Chemical Society **77(6)**: 1489–1494.

[84] Vavrusova M, Munk MB and Skibsted LH (2013) Aqueous solubility of calcium L–lactate, calcium D–gluconate, and calcium D–lactobionate: importance of complex formation for solubility increase by hydroxycarboxylate mixtures. Journal of Agricultural and Food Chemistry **61(34)**: 8207–8214.

[85] Cicek V (2012) Characterization studies of mild steel alloy substrate surfaces treated by oxyanion esters of α-hydroxy acids and their salts. International Journal of Chemical Science and Technology **2(3)**: 244–260.

[86] Marklund E, Öhman L-O and Sjöberg S (1989) Equilibrium and structural studies of silicon(IV) and aluminium(III) in aqueous solution. 20. Composition and stability of aluminium complexes with propionic acid and acetic acid. Acta Chemica Scandanavica **43**: 641–646.

[87] Wise DL, Trantolo DJ, Lewandrowski K-U, Gresser JD and Cattaneo MV (2000) Biomaterials Engineering and Devices: Human Applications: Volume 2. Orthopedic, Dental, and Bone Graft Applications. Springer, Dordrecht, Netherlands.

[88] Young Hee K (1967) Basic iron (III) alkanoate complexes. Masters Thesis—Paper 7135. Missouri University of Science and Technology, Rolla, MO, USA.

[89] Herdtweck E, Kornprobst T, Sieber R, Straver L and Plank J (2011) Crystal structure, synthesis, and properties of tri-calcium di-citrate tetra-hydrate $(Ca_3(C_6H_5O_7)_2(H_2O)_2).2H_2O$. Zeitschrift für Anorganische und Allgemeine Chemie **627**: 655–659.

[90] Sheldrick B (1974) Calcium hydrogen citrate trihydrate. Acta Crystallographica B **30**: 2056–2057.

[91] Froment DH, Buddington B, Miller NL and Alfrey AC (1989) Effect of solubility on the gastrointestinal absorption of aluminum from various aluminum compounds in the rat. Journal of Laboratory and Clinical Medicine **114**: 237–242.

[92] Dyer T (2016) Influence of cement type on resistance to organic acids. Magazine of Concrete Research, in press.

[93] Carde C and François R (1997) Effect of the leaching of calcium hydroxide from cement paste on mechanical and physical properties. Cement and Concrete Research **27(4)**: 539–550.

[94] Parkhurst DL and Appelo CAJ (2013) Description of input and examples for PHREEQC version 3—A computer program for speciation, batch-reaction, one-dimensional transport, and inverse geochemical calculations. In U.S. Geological Survey Techniques and Methods, Book 6. US Geological Survey, Denver, CO, USA, Chapter A43. See http://pubs.usgs.gov/tm/06/a43 (accessed 08/12/2016).

Chapter 3

Bacterial Biodeterioration

3.1 Introduction

Bacteria are single-celled organisms falling within the Prokaryota grouping. Prokaryote cells differ from the cells of other living organisms in their lack of a membrane-bound nucleus, membrane-bound organelles and mitochondria. As discussed in Chapter 1, the Prokaryote grouping has been subdivided into Archaea and Bacteria, but for the purpose of this book the two can be discussed together.

Bacteria comprise the largest proportion of biomass on the planet, and their presence on the Earth's surface (and to some distance beneath it) is practically ubiquitous. Indeed, bacteria have been found to be capable of existing in extremely hostile conditions.

The deterioration of concrete in contact with bacterial communities is principally through chemical reactions between cement and aggregate and substances produced by these organisms. However, it is of benefit to discuss the metabolism and reproduction of these organisms first, since these aspects have important implications with regards to interactions between bacteria and concrete.

3.2 Bacterial Metabolism

One means of categorizing bacteria is in terms of the manner in which they obtain the substances that they need to live and reproduce. These substances are:

- Carbon: carbon is required to manufacture the mass required for growth and division of bacteria cells.
- Reducing equivalents: these are substances capable of being used for the transfer of electrons through redox reactions. These are used by the organism either in processes used to store energy, or in the synthesis of chemical compounds in the cell.

In addition, cells require a source of energy. The manner in which the chemical and energetic requirements of bacteria can be met is discussed below.

Bacteria may source carbon *heterotrophically* or *autotrophically*. Autotrophic bacteria obtain carbon from carbon dioxide in the atmosphere, whilst heterotrophic bacteria use organic compounds in the surrounding environment, obtaining energy at the same time. Some bacteria are capable of obtaining carbon through a combination of both means, and are known as *mixotrophic* bacteria.

Reducing equivalents can either be obtained from inorganic chemical compounds—*lithotrophism*—or, again, from organic compounds—*organotrophism*.

Energy may be obtained *phototrophically* from sunlight or *chemotrophically* through the oxidation of inorganic compounds. Some bacteria are able to harness energy through other mechanisms.

Any combination of these different mechanisms is possible, with the bacteria being identified through a naming scheme that combines each term into one word, e.g., *chemolithoheterotroph*.

The harnessing of energy depends on the type of organism and whether the environment in which the bacteria are present contains oxygen (aerobic) or does not (anaerobic). In all cases, however, the basic reaction is the same: a reduced organic compound (AH_2) or reduced inorganic compound—for instance, hydrogen sulphide (H_2S)—donate hydrogens to oxidized compounds which act as hydrogen acceptors. *Obligate aerobes* are organisms that require oxygen to carry out this task. The basic process for aerobic heterotrophic bacteria can be summarized as:

$$2AH_2 + 2O_2 \rightarrow CO_2 + 2H_2O + \text{energy.}$$

In the case of aerobic chemotrophic bacteria, an example reaction would be the following:

$$H_2S + 2O_2 \rightarrow SO_4^{2-} + 2H^+ + \text{energy.}$$

Obligate anaerobes are organisms that are killed under normal aerobic conditions. Metabolism for obligate anaerobic heterotrophic bacteria follows the scheme:

$$AH_2 + B \rightarrow BH_2 + A + \text{energy}$$

where A is a reduced organic compound and B is oxidized organic compound.

Obligate aerobes and anaerobes exist at two extremes with regards to the manner in which bacteria respond to oxygen. Between these extremes exist *facultative anaerobic*, *microaerophilic* and *aerotolerant* organisms. Heterotrophic bacteria which are facultative anaerobes can metabolise both anaerobically or aerobically. In anaerobic conditions, metabolism will often employ hydrogen acceptors such as nitrate or sulphate as a source of oxygen:

$$AH_2 + NO_3^- + H^+ \rightarrow AH + N_2 + CO_2 + H_2O + energy.$$

AH is an intermediate compound which is typically an organic acid or an alcohol, and such bacteria are referred to as 'acid forming'. Heterotrophic 'acid-splitting' bacteria are potentially capable of breaking intermediates down further, e.g.:

$$2AH + SO_4^{2-} + 2H^+ \rightarrow CH_4 + S^{2-} + 2CO_2 + energy.$$

Microaerophiles metabolise aerobically, but are unable to survive at normal atmospheric oxygen concentrations. Aerotolerant bacteria have anaerobic metabolisms, but are unaffected by oxygen concentration.

As well as carbon, bacteria require additional elements for the purpose of cell growth and division in smaller concentrations. These include nitrogen, which is used in the synthesis of amino acids, DNA and RNA. Additionally, synthesis of nucleic acids requires phosphate, but this element is also necessary for the synthesis of adenosine triphosphate (ATP) which is used for energy transfer. Other elements required by bacterial cells in still smaller quantities include potassium, sodium, calcium, iron, zinc and magnesium. Some bacteria require even smaller quantities of other elements (micronutrients) for more specialized purposes.

As we have seen, with certain bacteria under anaerobic conditions, organic compounds may be formed as metabolites. However, bacteria are also capable of releasing compounds for the purpose of dissolving minerals to release nutrients. Bacteria colonising siliceous rock surfaces have been found to weather the surface, seemingly to obtain phosphorous present in trace quantities [1]. Indeed, it has been proposed that the presence of this element is the sole reason for the bacterial colonization of such surfaces. The compounds released include organic acids and polysaccharides, both of which are capable of forming complexes with orthosilicate (SiO_4^{4-}) ions. *Bacillus mucilaginosus* has been found to excrete both organic acids (oxalic, citric and lactic) and polysaccharides, with consequent accelerated weathering of mica and clay [2]. The polysaccharide, besides keeping the bacteria attached to the mineral surface, also appears to play two additional roles. Firstly, it keeps organic acids in close proximity to the surface and prevents them from dispersing more widely, thus maintaining a high acid concentration [2, 3]. Secondly, it adsorbs SiO_4^{4-} ions as they entered solution, thus preventing the solution at the surface of the mineral from achieving a state of equilibrium with the solid material, hence permitting further dissolution [3]. Other organic acids produced by bacteria, and which form complexes with SiO_4^{4-} include oxalic acid and 2-ketogluconic acid [4, 5]. Additionally, as will be discussed later, polymeric materials produced by bacteria in the formation of biofilms can also contain groups—such as the catecholate groups—which form these complexes.

Dissolution of carbonate minerals for the purpose of harvesting nutrients has also been proposed [6], and evidence of this is backed up by experimental results [7]. Dissolution of miliolite, a rock principally comprising calcite, aragonite (both $CaCO_3$) and quartz (SiO_2), was found to be conducted largely by three different bacterial genera: *Bacillus*, *Staphylococcus* and *Promicromonosporaceae*. Analysis of the organic acids formed identified acetic, lactic and malic acids as those having the largest influence on calcium dissolution [8]. Other studies have also found that the enzyme carbonic anhydrase by bacteria also has the effect of significantly increasing the dissolution of carbonate minerals [9].

3.3 Growth of Bacterial Communities

Bacteria reproduce asexually through binary fission: a cell divides into two cells which resemble the original. The manner in which a population of an isolated species of bacteria grows is described by the Monod equation [10], a version of which is:

$$\frac{dX}{dt} = \frac{\mu_{max}SX}{K_s - S}$$

where X =   the mass of bacteria;
   t =   time;
   μ_{max} =   the maximum specific growth rate, time^{-1};
   S =   concentration of the growth limiting substrate, mass/ volume; and
   K_s =   half saturation coefficient, mass/volume.

The growth limiting substrate is a nutrient required by the bacteria which is present in sufficiently low quantities that its use in metabolism will start to limit its availability, and hence limit growth. The specific growth rate (μ) is defined by the equation:

$$\mu = \frac{\left(\frac{dX}{dt}\right)}{X}$$

and the *maximum* specific growth rate is the specific growth rate at a concentration of substrate at or above saturation. The saturation constant is the substrate concentration required for a specific growth rate which is half that of the maximal rate.

Thus, growth is dependent on the population of bacteria and the concentration of the growth limiting substrate. The concentration of substrate will decline in a manner described by the equation

$$\frac{dS}{dt} = \frac{\mu_{max}SX}{Y(K_s - S)}$$

Where Y is a parameter known as the growth yield. The mass of bacteria will change in accordance with the plot shown in Figure 3.1. The initial

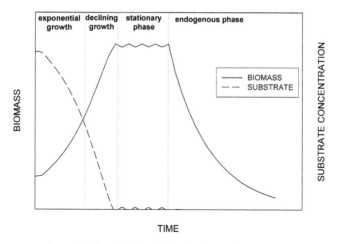

Figure 3.1 Growth of an isolated culture of bacteria [11].

process of growth is *exponential* in nature where the substrate is present in relative abundance, and the rate of growth is consequently limited only by the bacteria's capacity to employ it. As the substrate starts to become scarcer, its limiting effect begins to manifest itself and growth enters a *declining* phase where the rate of growth gradually declines.

Throughout the process of growth bacteria will be dying, and a *stationary phase* is eventually entered in which the rate of reproduction equals the death rate. The death of cells leads to the process of *lysis* where the membrane of the cells breaks down and the contents is released into the solution, potentially allowing the stationary phase to continue for longer. In some cases, this may lead to some of the substrate being released back into solution. A point is reached where the levels of substrate are so low that the rate of cells dying exceeds that of reproduction and the mass of bacteria declines in an *endogenous* phase.

A number of bacteria are capable of spore formation. Spores are reproductive units from which a bacterium can grow. Spore formation is not the primary means of reproduction of such bacteria. Instead, it provides a means for bacteria to survive hostile conditions: spores are not destroyed by dry conditions or low temperatures. Thus, spore formation acts as a mechanism by which a bacterial community can re-establish itself after periods of cold weather or drought.

Bacteria seldom occupy a particular environment as a single species, although there is likely to be a dominant species which is dictated by the nutrients and environmental conditions present. However, the metabolic processes of the dominant strain and the process of lysis will produce substances which can be used as nutrients by other species of bacteria, and organisms from other domains.

Many bacteria display a tendency to colonise surfaces, and the formation of *biofilms* plays an important role in this process. Biofilms are formed from biopolymers produced by the bacteria. These biopolymers largely consist of proteins and polysaccharides and are collectively known as extracellular polymeric substances (EPS). These polymers are initially formed by the bacteria to attach themselves, but continued formation leads to the formation of a film which links individual bacteria to neighbouring cells, encases the cells and provides a degree of 3-dimensionality to the colony, since it permits the formation of structures in which bacteria are stacked on top of each other.

As well as playing a structural role, biofilms can carry out a number of other functions. In many cases biofilms contain channels to provide microorganisms beneath the surface with nutrients or oxygen [12]. Conversely, biofilms can act as a barrier to atmospheric gases such as oxygen [13], or to sunlight [14]. It would also appear that the polysaccharide component of biofilms acts as a store of carbon which can be accessed during periods when nutrients would otherwise be in short supply [15]. Many biofilm molecules possess functional groups capable of forming bonds with metal cations in solution [16]. This mechanism has been suggested as a means of removing dissolved heavy metal ions from the local environment and thus avoiding the toxic effect these substances would have on bacteria [17].

The ability of biofilm compounds to form complexes with metal ions can allow biofilms to either accelerate the dissolution of calcium from surfaces or to act as the nucleation sites for the precipitation of calcium carbonate ($CaCO_3$). The first of these processes will be discussed in greater detail later on in this chapter. Biopolymer-induced precipitation of $CaCO_3$ can lead to the formation of either crystalline calcium carbonate or amorphous precipitates. It is believed that amorphous $CaCO_3$ precipitated in this way contains polysaccharide molecules incorporated into the structure [18]. It has also been proposed that enzymes released by the bacteria into the biofilm play a role in controlling this process in terms of location and possibly morphology [19].

A large proportion of bacteria are capable of movement. The most common form of motility is through the use of flagella. Flagella are rotary structures extending out of the outer membrane of the cell. When rotating, the flagella act as propellers, driving the bacteria through the surrounding fluid [20]. Other cellular architectures may be used for motion. These include Type IV pili, junctional pore complexes, ratchet structures and contractimg cytoskeletons. Type IV pili, are fibres projecting from the cell membrane which can be used to move bacteria across a surface through a series of jerks. This process is generally known as 'gliding'. Junctional pore complexes are also used as a means of gliding. These pores act as nozzles which permit motility through the extrusion of slime. Ratchet structures function through interaction between proteins in the cytoplasm of the cell and those in the outer membrane, leading to a movement of the outer membrane not unlike that of caterpillar tracks on vehicles.

The motion of bacteria is directed by local stimuli including concentrations of dissolved chemical species (chemotaxis), light (phototaxis) and oxygen (aerotaxis). Motion may be directed towards or away from such stimuli, depending on the nature of the bacteria.

3.4 Concrete as an Environment for Bacterial Life

In evaluating the surface of concrete as a habitat for bacteria, it becomes evident that, whilst many aspects of concrete offer a welcoming environment, there are other aspects which present a challenge to colonization. One of the more extreme characteristics of concrete—at least in its early life—is the very high pH that water takes on when in contact with its surface or when present within its pores. The physical nature of the surface is another aspect. Finally, bacteria require various nutrients, only some of which are present in concrete itself, and often in relatively small quantities. These factors are discussed below.

3.4.1 pH

High and low pH conditions are toxic to most bacteria. Whilst some species can survive under very extreme pH conditions, growth is fundamental to the survival of these organisms. Most bacteria are *neutrophilic*—they grow at optimal rates around a pH of 7. A small number of bacteria are able to survive at pHs as low as 1, and are referred to as being *acidophilic*. *Alkaliphilic* bacteria are capable of growth in the range of between 8.5 and just over 11, but their optimal rate of growth is less than this upper limit [21]. The pore solution of relatively young concrete will typically be in the range of pH 11–13, meaning that a new concrete surface is unlikely to support bacterial growth of any kind.

A number of different processes can occur in concrete which will act to reduce pH at the concrete surface. Probably the most common of these reactions is carbonation. This reaction, between carbon dioxide gas and cement hydration products in the presence of water, leads to the conversion of portlandite ($Ca(OH)_2$) to calcium carbonate:

$$CO_2 + H_2O \rightarrow H_2CO_3$$

$$Ca(OH)_2 + H_2CO_3 \rightarrow CaCO_3 + 2H_2O.$$

The reaction occurs at an optimal rate at a relative humidity of around 55%. Portlandite is only slightly soluble, but is the main source of hydroxide ions in hydrated cement. Thus, the reaction leads to a drop in pH to around

9. The reaction is diffusion-driven, with cement at the immediate surface being neutralised first and a 'carbonation front' progressing to greater depths, at an increasingly reduced rate (Figure 3.2).

　　When concrete is in contact with water, portlandite will be gradually leached out (see Chapter 2) also leading to a drop in pH, albeit over relatively long periods of time. Where acidic substances are dissolved in the water, this process is much quicker. Since acids are often formed by the metabolic processes of bacteria, and the presence of bacteria (or other microbes) near to a concrete surface may well render it suitable for colonization.

　　Concrete in contact with seawater has been found to form brucite $(Mg(OH)_2)$ at its surface [23] in place of portlandite. Brucite is considerably less soluble than portlandite, and so a drop in pH is again observed.

3.4.2　Concrete as a physical habitat for bacteria

The surface of concrete is relatively rough. Roughness can be measured in a number of ways, but one of the most common means of expressing it is in terms of R_a, the average vertical distance from the mean line of a surface measured as a surface is traversed. When such measurements are conducted on concrete, values in the order of 100s of µm are typically obtained [24]. Where surfaces have been mechanically ground after hardening, this may be an order of magnitude less [24, 25]. It might be expected that a rough surface would present an easier foundation for the establishment of bacteria populations. Studies with a focus on biomedical materials have found no clear relationship between surface roughness and growth of bacterial

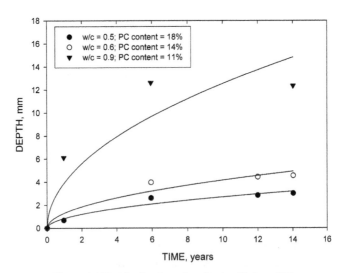

Figure 3.2 Depth of carbonation front with time [22].

communities [26, 27, 28, 29]. This is also reflected in the results of studies examining the relationship between the development of populations of cyanobacteria (and other organisms) on concrete surfaces, where no clear relationship was identified [30, 31]. However, the surface porosity of concrete does appear to influence the rate at which it is populated by bacteria, with a higher volume of porosity (via a higher water/cement ratio) leading to higher rates of colonization (Figure 3.3) [30, 31, 32]. The reason for this is unclear. Whilst it is possible that the pores provide features on the surface for bacteria to become lodged, it is most likely that the porosity simply provides a reservoir for moisture, which the bacteria need.

The previous discussion identified that many of the molecules produced by bacteria in the formation of biofilms possess functional groups capable of forming bonds with metal ions. This has two important implications from the perspective of bacterial attachment to concrete surfaces. Firstly, it is likely that this ability may well provide a mechanism for the formation of strong chemical bonds between biofilms and cement hydration products at the surface. One study has observed the formation of covalent bonds between biofilms produced by *Pseudomonas aeruginosa* and TiO$_2$ surfaces [33]. The compound responsible was identified as pyoverdine, which is a peptide siderophore. Siderophores are molecules which are capable of chelating iron, but also capable of forming strong bonds with other metal ions. In the case of pyoverdine, this appears to be via catecholate functional groups. Whilst this species has not been identified as one which is involved in deterioration of concrete, molecules with catecholate groups and similar

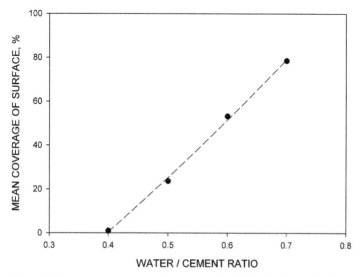

Figure 3.3 Coverage by a combination of cyanobacteria and algae of Portland cement mortar surfaces after 8 weeks in conditions favourable to growth [32].

are produced by many bacteria, making the formation of strong bonds with metal oxides in concrete a very likely possibility. It is also believed that divalent cations, such as calcium, can act as bridges between polysaccharide molecules, rendering biofilms stronger [34].

So far discussion of colonization of concrete by bacteria has been limited to surfaces. However, the fact that concrete is a porous material means that the possibility of bacteria penetrating further into the material exists. Concrete pores cover a wide range of diameters from nanometers to hundreds of micrometers. The pores in concrete take two forms: capillary pores and gel pores.

Gel pores are present in the CSH gel produced by the hydration of cement and have dimensions of between 0.5 and 10 nm. Capillary pores are the remnants of water-filled space between cement grains from when the concrete was in its fresh state. Much of the original space will be occupied by cement hydration products in hardened concrete, but what remains forms a network of pores, and has dimensions in the 10 nm to 10 μm range.

Bacteria range widely in size between species, generally within the range 0.5–1.0 μm [35]. This means that for many bacteria, much of the porosity—and certainly all of the gel porosity—is inaccessible. It is often tempting to imagine porosity in concrete as a series of tunnels of circular cross-section and uniform diameter running from one end of the material to another. The reality is much more complex—porosity consists of a highly interconnected network with high constrictivity (i.e., rapid change in diameter over a length of porosity). Taking this into account, cement porosity should be even less accessible to bacteria.

However, bacteria *are* able to penetrate beneath the surface of concrete, albeit gradually. One reason for this is that, as the concrete deteriorates, pore sizes will increase as their surfaces are dissolved. Additionally, cracks may form, leading to the interior becoming even more accessible. One study has examined bacterial communities in the deteriorated layers of asbestos cement pipes used for the transport of drinking water for over 50 years [36]. The researchers identified four layers—an outer layer where significant deterioration had occurred, two intermediate layers, and a deterioration front beyond which the cement was essentially unaffected. The bacteria were characterised in terms of their grouping (slime-forming, iron related, heterotrophic, acid-producing and sulphate-reducing) and populations of bacteria in each layer were estimated (Table 3.1). The most populous layer was the innermost layer, which contained mainly slime-forming bacteria.

3.4.3 Nutrients

There are usually no organic compounds present in concrete, other than very small quantities of chemical admixtures, formwork release agents and possibly cement grinding aids. Inorganic carbon may be present as

Table 3.1 Estimated populations of different bacteria in the deteriorated layers within an asbestos cement pipe used for carrying drinking water for a period of 52 years [11]. Total thickness of deteriorated zone was between 3.1 and 4.6 mm, but thickness of individual layers is not specified.

Layer	Estimated Population of Bacterial Group, active cells/g				
	Slime-forming	Iron-related	Heterotrophic	Acid-producing	Sulphate-reducing
Deterioration front	5,552,100	27,900	Nd	Nd	Nd
Intermediate layer I	2,150,500	34,500	115,000	Nd	Nd
Intermediate layer II	18,333	40,740	8,633	29,876	Nd
Outer layer	28,840	35,560	75,460	Nd	Nd

carbonate minerals, either as aggregate, included as a cement component, or calcium carbonate minerals deriving from carbonation of the cement matrix (Section 3.4.1). These carbonate minerals will partially dissolve in water. The dissolved inorganic carbon will either be in the form of carbonic acid (H_2CO_3), bicarbonate ions (HCO_3^-) or carbonate ions (CO_3^{2-}), with the dominant species dependent on pH (Figure 3.4). This inorganic carbon is available to autotrophic bacteria.

Some autotrophic bacteria struggle to obtain inorganic carbon under neutral pH conditions. This is because such bacteria require carbon to be in the form of CO_2, principally because it has a neutral charge and can therefore pass easily through their lipid membranes, whereas bicarbonate is the dominant species in these conditions. However, a very large proportion

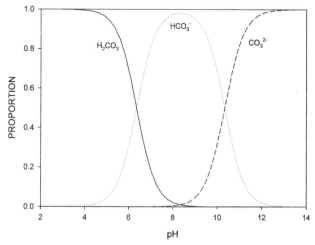

Figure 3.4 Change in the concentrations of Predominance of carbonic acid (H_2CO_3), bicarbonate (HCO_3^-) and carbonate (CO_3^{2-}) in water in contact with calcite.

of the bacteria found on external building surfaces are cyanobacteria which possess transporter regions in their cytoplasmic membrane which provide a means of bringing bicarbonate ions into the cell's interior [37].

Other autotrophic bacteria can convert carbonate ions to carbonic acid and subsequently to water and CO_2 by using enzymes extracellularly to catalyse the reaction [38]:

$$HCO_3^- + H^+ \rightleftharpoons H_2CO_3 \rightleftharpoons H_2O + CO_2.$$

Thus, concrete (or at least carbonated concrete or concrete containing carbonaceous aggregates) can provide some bacteria with a source of carbon.

In examining the metabolic processes of bacteria, we have seen that both nitrogen and sulphur can play roles in how bacteria obtain energy. Nitrogen compounds are typically present in very small quantities in concrete.

As will be seen in subsequent sections one group of bacteria—the sulphate-reducing bacteria—are capable of utilising sulphate as a source of oxygen. Sulphate in concrete will come mainly from calcium sulphate compounds (usually gypsum) added to Portland cement to control setting times. The sulphate content of cement is limited by many industrial standards. EN 197-1 [39] limits it to below 5% by mass as SO_3, although this can be lower for certain cement types. Assuming the highest cement content that would be encountered in a conventional concrete mix would be around 800 kg/m³ (a very rare situation), the maximum possible sulphate content—using the highest sulphate content for Portland cement—would be around 2% by mass or 0.2 mol/kg. Most typical concrete would have a sulphate content of a third of this, or less. With the additional issue of availability of this substance, given that access to nutrients within concrete is largely via the surface, this represents a very limited supply.

Other compounds which are required in smaller quantities by bacteria are usually present to some extent in concrete. Phosphate (P_2O_5) may be present in both cement or aggregate. Phosphate has a detrimental influence on strength development in Portland cement [40], although most industrial standards do not actively limit it. In recent years the possibility for higher phosphate levels in cement has been realized through the combustion of waste fuels [41]. A recent survey of Slovakian cement output, where use of such fuels was known to occur found the maximum level of phosphate in CEM I (Portland) cement to be 3.22% by mass, with commercial cements containing slag, fly ash and possibly other materials displaying lower levels [42].

Industrial by-products used alongside Portland cement may also contain phosphate. The production of fly ash in coal-fired power stations can sometimes involve 'co-combustion' where other waste fuels are burnt alongside the coal. As for Portland cement, where the wastes contain phosphorous (for instance, sewage sludge) phosphate levels in the ash may be relatively high. For this reason the current European standard for fly ash for use in concrete limits total P_2O_5 levels to less than 5.0% by mass, and soluble phosphate to 100 mg/kg [43].

The alkali metals potassium and sodium are present in all concrete constituents. Regardless of the issue of biodeterioration, the alkali content of concrete is of concern from a durability perspective, since certain susceptible aggregates may undergo alkali-aggregate reactions at high alkali contents. Alkalis in Portland cement are usually of greatest concern, since they are typically present in highly soluble sulphate salt form on the surface of cement grains, usually converting to alkali hydroxides during hydration. In cement, alkali content is expressed as the equivalent of sodium oxide (Na_2O), which is calculated using the equation:

$$Na_2O_{eq} = Na_2O + 0.658K_2O$$

A study of Portland cement manufactured in the USA and Canada in 2004 found the average level of alkalis to be 0.70% Na_2O_{eq}, for Type I cement, with a maximum of 1.20. Other types of cement displayed slightly lower averages. Fly ash and silica fume typically possess alkali levels somewhat higher than this [44, 45, 46], whilst slag generally contains lower quantities [47].

Alkalis in aggregate are often in a much less soluble form, frequently present in silicate minerals such as feldspars, mica and clay-forming minerals. However, these can gradually be released under the high pH conditions of concrete [48] and, as discussed previously, bacteria can potentially accelerate their release through the production of organic acids and/or complex-forming compounds. Alkali may also be introduced into concrete dissolved in mix water.

Iron is present in Portland cement in relatively large quantities (typically around 2–4% by mass as Fe_2O_3). In the hydrated state it is present largely as the AFm and AFt phases and as a substituted ion in CSH gel. As the pH of concrete is reduced, the AFm and AFt phases will decompose (see Chapter 2). However, the iron will remain in a relatively insoluble form as ferrihydrite, until much lower pH conditions are reached. The presence of organic acids capable of forming complexes with iron may increase the pH at which iron becomes soluble. Many aggregates will contain quantities of iron, although commonly this is present as hematite (Fe_2O_3) or possibly magnetite (Fe_3O_4) which are essentially insoluble. Again, complexation by organic acids and other compounds may render iron more soluble.

Magnesium is also present in Portland cement, although its levels are again limited by industrial standards. EN 197-1 limits MgO in Portland cement clinker to 5.0% by mass, whilst ASTM C150 limits it to 6.0% in all cements covered by the standard [49]. The European standard for fly ash in concrete limits MgO to less than 4.0% by mass [43], whilst the equivalent standard for slag limits it to 18% [50].

Calcium is present in large quantities in concrete, certainly from cement, but also potentially from aggregate. It has been observed that cyanobacteria have a tendency to preferentially grow on surfaces containing calcareous surfaces, with colonies of bacteria present on mortar between siliceous stone [51, 52]. Whilst this presents the possibility that either the presence of calcium or carbonate, or both, is the reason for this, a study of exclusively calcareous building surfaces found that cyanobacteria preferred cement and mortar surfaces to those of limestone, which is also rich in calcium [53]. It must be concluded, therefore, that it is the higher porosity of the cement and mortar which provides the favourable conditions, rather than the chemistry.

It can be concluded that whilst some of the nutrient required in small quantities by bacteria may be present in concrete, a number of key nutrients are in short supply, specifically nitrogen and organic carbon. This, however, does not mean that bacteria cannot successfully colonise concrete surfaces. Instead, it means that the types of bacteria that can colonise a concrete surface are limited to those which are able to obtain the absent nutrients from other parts of the surrounding environment.

3.5 Deterioration of Concrete

Bacterial deterioration of concrete is primarily the result of chemical attack deriving from substances produced by the metabolic processes of bacteria, and it is three such forms of this which will be examined first. However, the presence of bacterial communities on concrete surfaces also has aesthetic implications, and so this aspect is also explored.

3.5.1 Attack by sulphate reducing and sulphur-oxidising bacteria

By far the most documented concrete deterioration process involving bacteria is the attack of sewage pipes and related structures by biogenic sulphuric acid. This form of attack follows a relatively convoluted route involving many aspects of chemistry and physical conditions which are unique to such environments. Moreover, it requires two types of bacteria whose metabolisms function in very different ways. The processes involved are summarized in Figure 3.5, and described in more detail below.

The first type of bacteria involved in the process are a subset of the group known as sulphate reducing bacteria. The main species encountered in sewer pipes is *Desulfovibrio desulfuricans* [54]. These bacteria are obligate anaerobes and can only exist in accumulations of biofilm and sludge which occur on sewer pipe surfaces just above the waterline where oxygen levels are sufficiently low. They obtain their energy by oxidising organic compounds (AH in Figure 3.5 and principally lactic acid) and hydrogen gas, and obtain the oxygen required for this by reducing sulphate ions

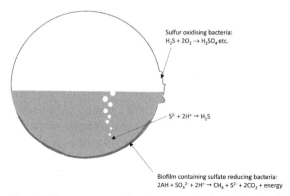

Figure 3.5 Processes occurring in a sewer pipe leading to attack of concrete by sulphuric acid.

(SO_4^{2-}) dissolved in sewage. The resulting sulphide ion (S^{2-}) rapidly reacts with hydrogen dissolved in the sewage to give hydrogen sulphide (H_2S).

The solubility of sulphide in water is dependent on pH. This is because H_2S undergoes dissociation at higher pHs to give bisulphide (HS^-) ions and a variety of sulphide ions at even higher pHs, increasing the capacity of the water to contain dissolved sulphide. The dissociation constants of H_2S at different temperatures are given in Table 3.2.

Where the pH of sewage is above 7, its capacity for dissolved sulphide is relatively high. However, where the pH is lower—which is often the case for sewage—the capacity is limited and H_2S escapes as gas into the air space above the sewage in the pipe. The reason for the low pH is also tied to microbial activity—organic matter in the sewage is being broken down into organic acid intermediates, which will also act as a source of carbon to the sulphate reducing bacteria [57]. It can also be seen from Table 3.2 that higher temperatures will also encourage the release of H_2S.

The H_2S above the sewage now experiences a different chemical environment: at the surface of the concrete pipe is usually a thin layer of condensed water which, being in contact with concrete has a relatively high pH.

Table 3.2 Dissociation constants of H_2S at different temperatures [55, 56].

Temperature, °C	Dissociation Constant (pka)						
	$H_2S \rightleftharpoons HS^- + H^+$	Sulphide Ions					
		$HS^- \rightleftharpoons S^{2-} + H^+$	S_2^{2-}	S_3^{2-}	S_4^{2-}	S_5^{2-}	S_6^{2-}
25	7.00	17.30	11.78	10.77	9.96	9.37	9.88
71	6.55	–	–	–	–	–	–
93	6.46	–	–	–	–	–	–
104	6.44	–	–	–	–	–	–

This means that the layer of moisture can dissolve relatively large quantities of H_2S. Oxygen in the atmosphere above the sewage can react with the H_2S to form a range of different solid compounds including elemental sulphur, thiosulfates and polysulphates (tetrathionates, pentathionates, etc.) [58].

It is at this point that the second group of bacteria play a role. These are the species *Thiobacillus* which are also present at the concrete surface. *Thiobacillus* is one of a group of species known as colourless sulphur bacteria, or sulphur-oxidising bacteria. These organisms are chemotrophs and obtain their energy by oxidising inorganic compounds. Many reactions can be used for this purpose, as outlined in Table 3.3. The products of most of these reactions is either sulphuric acid (H_2SO_4) or compounds which can subsequently by used in reactions which yield sulphuric acid. Another common group of sulphur-bearing compounds in sewer pipes are the mercaptans—organic compounds with a thiol (–SH) group. These, however, do not appear to be usable by thiobacilli [60].

Bacteria identified in or on samples of concrete taken from corroded sewer pipes from three studies are shown in Table 3.4. Common to all these studies is *Thiobacillus thiooxidans*. Also included in the table are the pH ranges within which different *Thiobacillus* species can grow. It should be noted that *Thiobacillus thiooxidans* is only able to grow in conditions of low pH, and that none of the other species grow under pH conditions greater than 9 or 10. For this reason, it is necessary for the surface of the concrete, which would otherwise have a pH of 11 or more, to become less alkaline. The dissolution of acidic H_2S in the water layer will play a small role in the neutralization of the cement hydration products which are responsible for the high pH. However, the largest contribution is likely to come from carbonation—CO_2 concentrations in the airspace in the pipe are in most cases

Table 3.3 Reactions used by colourless sulphur-oxidising bacteria to obtain energy [61].

Compound	Reaction
H_2S	$H_2S + 2O_2 \rightarrow H_2SO_4$
	$2H_2S + O_2 \rightarrow 2S^0 + H_2O$
	$5H_2S + 8KNO_3 \rightarrow 4KSO_4 + H_2SO_4 + 4N_2 + 4H_2O$
Sulphur	$2S^0 + 3O_2 + 2H_2O \rightarrow 2H_2SO_4$
	$5S^0 + 6KNO_3 + 2H_2O \rightarrow 3KSO_4 + 2H_2SO_4 + 3N_2$
Sodium thiosulphate	$Na_2S_2O_3 + 2O_2 + H_2O \rightarrow Na_2SO_4 + H_2SO_4$
	$4Na_2S_2O_3 + O_2 + 2H_2O \rightarrow 2Na_2S_4O_6 + 4NaOH$
Sodium tetrathionate	$2Na_2S_4O_6 + 7O_2 + 6H_2O \rightarrow 2Na_2SO_4 + 6H_2SO_4$
Potassium thiocyanate	$2KSCN + 4O_2 + 4H_2O \rightarrow (NH_4)SO_4 + K_2SO_4 + 2CO_2$

Table 3.4 Bacteria identified from samples of corroded concrete from sewer pipes.

Organism	ph for Growth [59]	Carbon Fixation [59]	Reference
Thiobacillus thiooxidans	0.5–4.0	Autotroph	[62, 63, 64]
Thiobacillus neapolitanus	4.0–9.0	Autotroph	[63]
Thiobacillus intermedius	1.7–9.0	Mixotrophic	[63]
Thiobacillus novellus	5.0–9.2	Mixotrophic	[63]
Thiobacillus ferrooxidans	1.5–2.5	Autotroph	[64]
Bacillus			[62]
Ochrobactrum anthropi			[62]
Microbacterium			[62]
Pseudomonas			[63]

likely to be high, due to microbial activity, leading to rates of carbonation considerably faster than under normal atmospheric conditions.

When sulphuric acid comes into contact with concrete made using Portland cement, two processes detrimental to the integrity of the material occur simultaneously. The first of these is acidolysis, whilst the second is sulphate attack. Because both forms of attack influence each other, it is necessary to discuss them as a single process.

When sulphuric acid comes into contact with the hydration products of Portland cement it will lead to ions from these products becoming dissolved in solution. For instance, in the case of one of the main components of hydrated Portland cement, Portlandite ($Ca(OH)_2$):

$$H_2SO_4 + Ca(OH)_2 \rightarrow Ca^{2+} + SO_4^{2-} + 2H_2O$$

The resulting calcium and sulphate ions will, under the right conditions, precipitate as gypsum ($CaSO_4.2H_2O$). However, where calcium aluminate hydrates are present in the cement, such as monosulphate ($3CaO.Al_2O_3.CaSO_4.12H_2O$), a different process may occur if the pH is sufficiently high:

$$3CaO.Al_2O_3.CaSO_4.12H_2O + 2Ca^{2+} + 2SO_4^{2-} + 20H_2O \rightarrow 3CaO.Al_2O_3(CaSO_4)_3.32H_2O$$

The product of this reaction is ettringite. The precipitation of both these compounds with respect to solution pH has been examined in Chapter 2 in the section covering sulphuric acid. These two compounds are stable at high pHs and have high molar volumes. Thus, where the pH is high these compounds precipitate leading to expansion and cracking of the cement matrix. However, sulphuric acid continues to ingress below the concrete surface and as this occurs and the reactions continue, the $Ca(OH)_2$, which provides the main source of OH^- ions, is neutralised and the pH drops. As pH drops, gypsum and ettringite become soluble and CSH gel begins to decalcify. The overall effect of this is that considerable mass is lost from the concrete surface and its strength is substantially reduced.

The effect of the above process is shown in Figure 3.6, which shows the results of a geochemical model in which hydrated cement paste is brought into contact with sulphuric acid. It can be seen that the outer part of the cement paste is completely dissolved, followed by a layer in which decalcified CSH is present. Beyond this layer are deposits of small quantities of ettringite and substantial quantities of gypsum, followed by unaffected hydrated cement. These modelled results match well with the results of powder X-ray diffraction carried out on the acid-affected layers of concrete specimens [65]. Under conventional sulphate attack, it would be not unusual

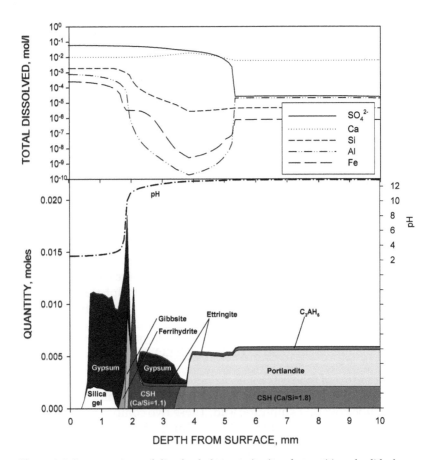

Figure 3.6 Concentrations of dissolved elements (top) and quantities of solid phases obtained from geochemical modelling of sulphuric acid attack of hydrated Portland cement. Model conditions: acid concentration = 0.1 mol/l; volume of acid solution = 4 l; mass of cement 80 g; diffusion coefficient = 5×10^{-13} m^2/s.

to find higher quantities of ettringite. The dominance of gypsum under sulphuric acid attack derives from the stability of this mineral down to the relatively low pH conditions characteristic of acid attack.

An alternative mechanism has been proposed in which iron from cement in the acid-affected zone of the concrete (where pH is low) is dissolved (See Chapter 2), diffuses further into the concrete (where pH is higher) and precipitates as a thin layer of ferrihydrite [66]. This process is also evident in the geochemical modelling results in Figure 3.6. The precipitation of ferrihydrite is well-known as having the effect of creating expansive stress around corroding steel reinforcement, and it is proposed that the same effect is observed in this case. If this mechanism does occur, it is not unique to sulphuric acid attack, but would occur for any other process of acidolysis, which does not appear to be the case.

The cracking resulting from the precipitation of expansive products should permit the ingress of sulphuric acid at a faster rate. However, this appears to be offset to a large extent by the precipitation of the reaction products, leading to a reduction in porosity and a reduction in the diffusion coefficient of sulphate into the concrete in the zone where these products are present [64].

The effect of sulphuric acid exposure on concrete porosity is shown in Figure 3.7. There is a reduction in total porosity with most of this reduction occurring in the finer pore size range, which is in accordance with what

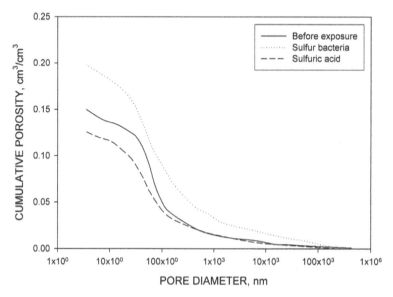

Figure 3.7 Cumulative pore size distributions obtained using mercury intrusion porosimetry for concrete specimens placed in a reactor containing sulphur oxidizing bacteria or submerged in a 0.15 mmol/l solution of sulphuric acid for 14 days [67].

might be expected from the precipitation of gypsum. Also shown in this figure is the pore size distribution obtained for concrete stored in a bioreactor containing sulphur oxidizing bacteria. It is notably different to that obtained for sulphuric acid exposure—there is an increase in total porosity, with a pore size distribution curve indicative of a widening of pores of all sizes. It should be stressed that these results were obtained using mercury intrusion porosimetry, which is unable to distinguish between cracks and pores. Nonetheless, these results suggest that a degree of caution should be employed before considering sulphuric acid exposure as a model for deterioration resulting from the activity of sulphur oxidizing bacteria.

The drop in pH that results as sulphuric acid ingress proceeds will of course make more and more of the concrete surface and inner volume of concrete favourable to colonisation by sulphur oxidising bacteria, and also a shift towards more acid-tolerant species. As the pH declines, there is a shift from an abundance of bacteria such as *Thiobacillus neapolitanus* towards the more acidophilic *Thiobacillus thiooxidans* [63].

The process of sulphuric acid attack leads to a loss in mass from the concrete, which consequently leads to a loss in cross-sectional area and strength. The sulphuric-acid environment to which concrete is exposed plays an important role in its rate of deterioration. The concentration of acid will obviously play an important role, with higher concentrations leading to lower pH conditions and faster rates of deterioration (Figure 3.8). pH values as low as 1 may be measured in concrete which has undergone attack by *Thiobacillus thiooxidans* [63].

Another important factor is whether the sulphuric acid is replenished, or whether a finite quantity of acid is present. In the latter case, the acid will gradually become depleted as it reacts with cement hydration products, and the deterioration rate will decline with time, whereas the rate of deterioration is steady where replenishment occurs. This is illustrated in Figure 3.9, which shows mass loss rates from concrete exposed to both types of regime. In the case of biogenic formation of sulphuric acid in sewer pipes, the conditions are such that a replenishing acidic environment is the most likely scenario, since the flow of sewage through the pipe will ensure a constant supply of nutrients to the system.

One study has examined mass loss from concrete suspended in a reactor containing simulated wastewater and various sulphur oxidising bacteria, with the atmosphere above the liquid steadily replenished with H_2S [70]. This led to a fairly uniform pH being established and the results of mass loss measurements (Figure 3.10) indicate a reasonably uniform rate of mass loss characteristic of a constant sulphuric acid concentration.

The decalcified and deteriorated surface tends to remain at least partially in place unless it is subject to mechanical disruption. It possesses very little strength and so is easily removed. Many studies examining the influence of sulphuric acid utilise periodic brushing of the concrete surface

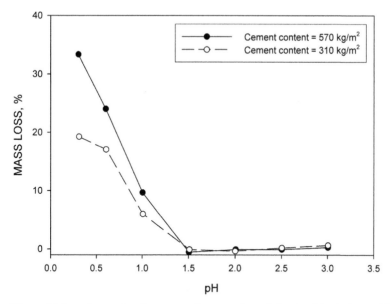

Figure 3.8 8-week mass loss from concrete exposed to sulphuric acid solutions with different concentrations (expressed in terms of the resulting pH). Specimens were not abraded [68].

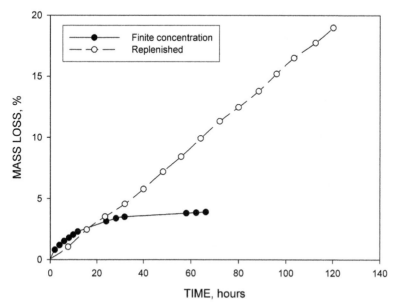

Figure 3.9 Mass-loss versus time plots for concrete specimens exposed to a finite concentration of sulphuric acid (pH 1.0) and one which is replenished every 8 hours [69].

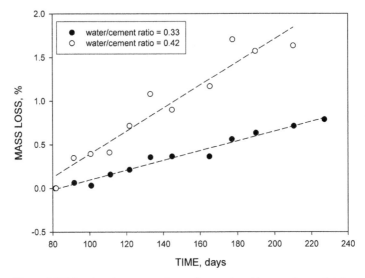

Figure 3.10 Mass loss from concrete specimens stored in a reactor containing simulated wastewater and sulphur oxidizing bacteria [70].

to mimic the effect of the flowing water in sewers. The extent to which this mimics reality is debatable, given that corrosion occurs above the water-line. However, this level will in fact vary with time and so corroded zones are likely to experience flowing water periodically. The rate of mass loss resulting from brushing is greater than that without brushing (Figure 3.11). Clearly the majority of this increase is simply the result of removal of material. However, it should also be remembered that the reaction products may protect the underlying concrete from further attack, and brushing will diminish this effect.

Concrete characteristics which influence the magnitude to which sulphuric acid attack causes deterioration include cement content, water/cement ratio, cement type and aggregate type.

Since sulphuric acid attack is largely a cement deterioration process, the cement content plays a major role in determining the response of concrete. As the cement content increases, it increases the availability of calcium, which is an essential component in the formation of gypsum and ettringite. Moreover, it is largely the cement which will be dissolved as a result of the process of acidolysis. Thus, as cement content increases, so too does the rate of deterioration (Figure 3.12).

The relationship between water/cement ratio and concrete durability under sulphuric acid attack runs counter to expectation. A reduced water/cement ratio yields higher strengths and reduced porosity, both of which provide greater resistance against many forms of deterioration. This appears

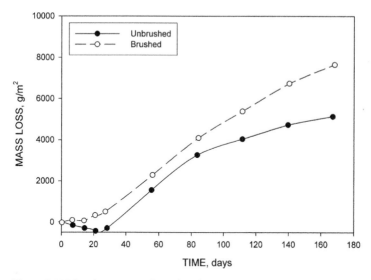

Figure 3.11 Mass-loss versus time plots for concrete specimens exposed to a pH 1.5 sulphuric acid solution with and without periodic abrasion using a brush [71].

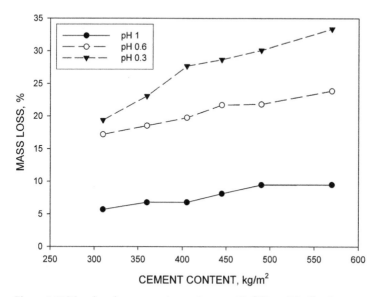

Figure 3.12 Mass loss from concrete specimens with different Portland cement contents after exposure to sulphuric acid solutions for a period of 60 days [68].

to not be the case for sulphuric acid, regardless of whether the concrete undergoes abrasion [72, 73, 74, 75]. This can probably be attributed to two factors. Firstly, the conventional means of achieving low water/cement ratios during concrete mix design is to increase the cement content, yielding the increase in vulnerability discussed above. Secondly, the increased porosity at high water/cement ratios may offer a larger volume of space within which the reaction products are able to precipitate leading to a delay in the onset of expansive forces. Moreover, the accumulation of reaction products in this way is likely to offset mass loss from fragmentation and acidolysis.

However, results obtained from sulphuric acid exposure experiments do not present the whole picture, and Figure 3.10 indicates that, where attack derives from sulphur bacteria, a low water/cement *does* reduce the rate of attack. This issue is discussed further in the chapter when measures to limit rates of deterioration are discussed.

Cementitious materials used in combination with Portland cement tend to provide enhanced resistance to sulphuric acid in comparison to concrete containing only Portland cement. This is likely to be for two reasons. Firstly, the cement fraction will possess a reduced calcium content, thus limiting the quantities of gypsum and ettringite formed. Secondly, many such materials have the effect of refining the pore structure of concrete either through fine particles occupying the space between coarser Portland cement particles, or through the formation of reaction products as a result of pozzolanic or latent hydraulic reactions. Table 3.5 summarises the findings of studies examining the performance of concrete containing ground granulated blastfurnace slag (GGBS), fly ash (FA), silica fume (SF), natural pozzolanas (NP) and metakaolin (MK) with regards to sulphuric acid resistance. GGBS appears to have only a minor influence over resistance, presumably due to its relatively high calcium content. Fly ash, whose presence will reduce calcium content considerably more, improves performance to a much greater extent (Figure 3.13). Whilst silica fume and metakaolin improve performance in most of the studies included in Table 3.5, this cannot be attributed to reduction in calcium content, since they are used at relatively low levels, and so refinement of porosity is likely to be the mechanism of enhanced resistance.

Calcium aluminate cements have also been found, in some instances, to display enhanced resistance to sulphuric acid. However, results vary considerably from study to study, possibly due to variation in composition of these materials. A field study in which calcium aluminate cement (CAC) was used in concrete mixtures in sewer pipes alongside Portland cement concrete pipes indicated considerably less deterioration from the CAC material when evaluated through inspection [69]. However, laboratory experiments conducted by the same researchers did not replicate these results. The same study also placed specimens made from sewer pipe

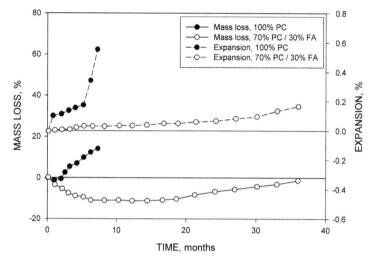

Figure 3.13 Mass loss and expansion of Portland cement and Portland cement/ fly ash concrete specimens exposed to a 2% sulphuric acid solution with periodic replenishment to maintain pH [78].

lining mixes made using both CAC and Portland cement in sewers and measured mass loss. The CAC lining materials performed better than their Portland cement equivalents. A similar study in which concrete specimens were suspended in sewers found similar results when CAC concrete was compared against Portland cement and Portland cement/GGBS blend concrete [86].

One reason for enhanced resistance of CAC concrete, where this is achieved, is most probably the lower calcium content of the material. The results of geochemical modelling of sulphuric attack of a typical hydrated calcium sulphoaluminate cement is shown in Figure 3.14. Comparing these results to those obtained for Portland cement (Figure 3.6) it is evident that the peak quantity of gypsum produced is lower for the calcium aluminate cement. However, a substantial quantity of this mineral is still present, and so only minor improvement in performance might be expected. It would also appear that the high aluminate content of the cement has the effect of suppressing bacterial growth: it has been shown that aluminium concentrations in excess of 350–600 mg/l stop growth of *Thiobacillus thiooxidans* [87]. This would certainly explain the relatively poor performance of CAC concrete when exposed to sulphuric acid solutions, compared to performance in the field.

The possibility of using limestone aggregates as a means of enhancing sulfuric acid resistance has been explored, on the basis that the calcium carbonate in these materials is capable of neutralising acids [76, 88]. The presence of limestone does tend to enhance resistance, although it

Table 3.5 Findings of research conducted to evaluate resistance of concrete containing pozzolanic and latent hydraulic materials to sulphuric acid attack. FA = fly ash; SF = silica fume; NP = natural pozzolana; MK = metakaolin.

Cement		Study [72]	[71]	[73]	[74]	[75]	[76]	[78]	[80]	[81]	[82]	[83]	[84]	[84]	[84]
GGBS	Level	Not stated, but 36–65%	50–70% with limestone												
	Acid Conc.	Alternating 50 mg/l NaSO$_4$ and pH 2 H$_2$SO$_4$	1%												
	Outcome	Slight improvement	Negligible change												
FA	Level			7.5–22.5% siliceous fly ash	33% siliceous fly ash, 7% SF	Optimum at 50% siliceous fly ash, 10% SF	10%–70% siliceous fly ash	13.5–50% siliceous fly ash							
	Acid Conc.			1%	1% replenished weekly	1% replenished weekly	2% with periodic additions	8.7% H$_2$SO$_4$							
	Outcome			Reduced	Enhanced	Enhanced	Enhanced	Slightly enhanced							
SF	Level			7.5–30%					7.5–30%	15%	5–15%				
	Acid Conc.			1%					1 and 5%	1%	2%				
	Outcome			Negligible change			Enhanced at > 5%		Enhanced	Enhanced	Enhanced				

NP	Level	13.5–50% Algerian natural pozzolana	40% 'true pozzolana'	28% trass + 6% calcareous fly ash
	Acid Conc.	8.7%	2:1 sulphuric/nitric, pH 3.5, periodic additions	0.5% to 1.2% + periodic additions
	Outcome	Slightly enhanced	Reduction	Reduction
MK	Level	7.5–30%	5–15%	
	Acid Conc.	1%	2%	
	Outcome	Negligible change	Enhanced	
Others	Level	2.5–20% corn cob ash		
	Acid Conc.	35%		
	Outcome	Enhanced up to 10% of cement		

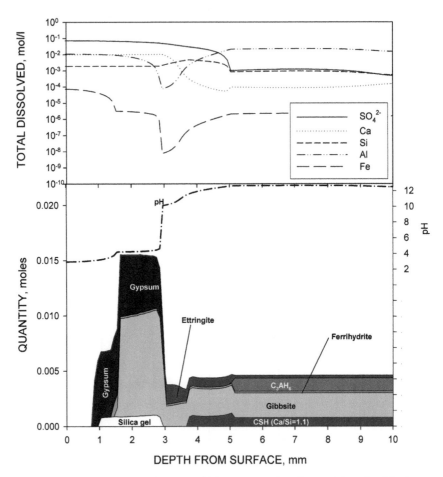

Figure 3.14 Concentrations of dissolved elements (top) and quantities of solid phases obtained from geochemical modelling of sulphuric acid attack of a typical hydrated calcium sulphoaluminate cement. Model conditions: acid concentration = 0.1 mol/l; volume of acid solution = 4 l; mass of cement 80 g; diffusion coefficient = 5×10^{-13} m^2/s.

seemingly only delays the process of deterioration (Figure 3.15)—it should be remembered that whilst the capacity to neutralise acids is enhanced, the calcium content of the concrete will be increased. Nonetheless, use in combination with pozzolanic materials and possibly other measures to improve resistance would appear to be a valid approach. When limestone powder is used as a cement filler component, resistance is not enhanced [79].

Significantly enhanced performance has been achieved where slow-cooled (and hence crystalline) blastfurnace slag fine aggregate has been used in place of siliceous sand. It has been proposed that this results from

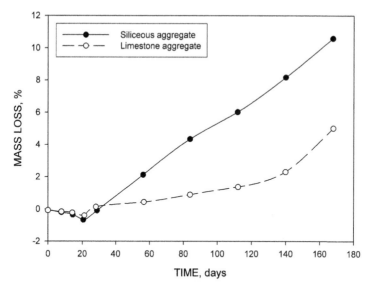

Figure 3.15 Mass loss from concrete specimens containing different aggregate exposed to a 1% sulphuric acid solution [76].

the formation of a denser layer of protective gypsum in the acid-affected zone. However, why this should be the case is not clear.

Synthetic aggregate made from calcium aluminate cement clinker has been shown to substantially enhance sulphuric acid resistance of concrete made using CAC, compared to dolomite aggregate [68]. It should be noted that the particle size of the two aggregates was also different, with the synthetic aggregate having finer particles.

Whilst the process of corrosion by sulphate reducing and sulphur oxidising bacteria is common in sewers, it is not exclusively limited to these environments. Deterioration of concrete by this process in outlet tunnels in two artificial lakes in Ohio, USA has been reported in the literature [88]. In these cases, sulphate levels in the water were high (> 300 mg/l) and attributed, at least for one lake, to strip-mining activities upstream.

3.5.2 Attack from nitrifying bacteria

Bacterial activity can also produce nitric acid. As with sulphuric acid formation, the process is relatively complex, involving a chain of interaction between various bacterial groups. The starting point of biogenic nitric acid attack is the compound urea ($CO(NH_2)_2$). Urea can be used by a group of bacteria known as urobacteria. Urobacteria hydrolyse urea for energy, producing ammonia (NH_3):

$$CO(NH_2)_2 + H_2O \rightarrow 2NH_3 + CO_2$$

A wide range of bacteria are capable of this process.

Ammonia is employed by a group of bacteria collectively known as ammonia-oxidising bacteria (see Table 3.6). These organisms are autotrophic (but are capable of mixotrophism) and grow optimally at temperatures of 25–30°C at pH 7.5–8.0 [90]. They obtain energy through oxidising ammonia in the following manner:

$$2NH_3 + O_2 \rightarrow 2NO_2^- + 3H^+ + 2e^-$$

The simultaneous formation of nitrite ions and protons means that nitrous acid (HNO_2) is formed. Nitrite ions (NO_2^-) are oxidised further by nitrite-oxidising bacteria (Table 3.7):

$$2NO_2^- + H_2O \rightarrow NO_3^- + 2H^+ + 2e^-$$

Again, because protons are produced, nitric acid (HNO_3) is effectively the product of this reaction.

The formation of nitric acid through bacterial activity has historically been a problem in water treatment and distribution systems, where conditions are suited to the organisms involved, and where quantities of ammonia are typically present. It is for this reason that these environments

Table 3.6 Ammonia-oxidising bacteria found in water distribution and treatment environments.

Genus	Species	Location	References
Nitrosomonas	–	Water distribution systems	[91]
	europaea	Biofilm bioreactors	[92, 94]
		Wastewater treatment plants	[97]
	oligotropha	Trickling filters	[93]
		Chloraminated water distribution systems	[96]
		Wastewater treatment plants	[97]
	communis	Trickling filters	[93]
		Wastewater treatment plants	[97]
	uraea	Chlorinated water distribution systems	[95]
'Nitrosococcus'	*mobilis*	Biofilm bioreactors	[92]
Nitrosospira	-	Activated sludge bioreactors	[94]
		Chloraminated water distribution systems	[96]
		Wastewater treatment plants	[97]
	briensis	Chlorinated water distribution systems	[95]
'Nitrosovibrio'		Sandstone	[98]
		Sandstone, concrete	[99]

Table 3.7 Nitrite-oxidising bacteria found in water distribution and treatment environments.

Genus	Species	Location	Reference
Nitrospira	–	Trickling filters	[93]
		Chloraminated water distribution systems	[96]
	moscoviensis	Chlorinated water distribution systems	[95]
Nitrobacter	–	Biofilm bioreactors	[92]
		Chloraminated water distribution systems	[96]
		Wastewater treatment plants	[97]
		Sandstone, concrete	[99]

feature heavily in Tables 3.6 and 3.7. However, a growing problem in recent history has been the growth in concentrations of atmospheric ammonia.

The rise in atmospheric ammonia concentrations is the result of human activities, with such sources including coal combustion, fertilizer use and livestock production. Ammonia released into the atmosphere will tend to react with pollution-derived acids in the atmosphere (sulphuric acid, nitric acid, etc.) to form ammonium salts. These salts will be precipitated either as particulates or dissolved in precipitation [97]. As a result, high levels of ammonia are frequently found in porous building materials with surfaces exposed to the exterior environment [98].

Most nitrifying bacteria are obligate lithoautotrophs (with the exception of *Nitrobacter*, which can grow heterotrophically [90]), and so no source of organic matter is required. They are also obligate aerobes, needing oxygen for the nitrifying reactions. They are, however, killed by visible blue light and long-wavelength UV light [100]. As a result, they form biofilms containing significant quantities of EPS to act as a barrier to light.

In sandstone (which is typically more porous), *Nitrosovibrio* have been found at depths of 30 cm below the surface [99]. Nitrifying bacteria are also capable of penetrating concrete: whilst the 'slime forming' bacteria found at some depth inside asbestos cement water pipes discussed in Section 3.4.2 are not specifically identified, it is likely that they are nitrifying bacteria.

When nitric acid is brought into contact with hardened Portland cement, the effect is more straightforward than that for sulphuric acid. The cement hydration products will undergo acidolysis, leaving a decalcified zone at the surface. The results of geochemical modelling of the process of nitric acid attack are shown in Figure 3.16.

The results from the model indicate why the water pipe study discussed in Section 3.4.2 found the most nitrifying bacteria at the interface between the unaffected cement and the deteriorated material. The production of nitric acid, if unchecked, will lead to a drop in pH which will eventually

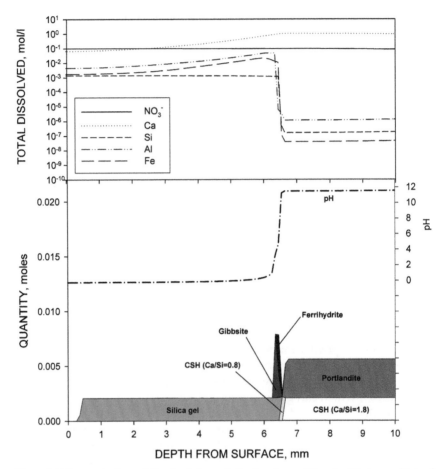

Figure 3.16 Concentrations of dissolved elements (top) and quantities of solid phases obtained from geochemical modelling of nitric acid attack of hydrated Portland cement. Model conditions: acid concentration = 0.1 mol/l; volume of acid solution = 4 l; mass of cement 80 g; diffusion coefficient = 5×10^{-13} m^2/s.

lead to a cessation in growth. However, where cement hydration products come into contact with nitric acid, neutralization will occur, e.g.:

$$Ca(OH)_2 + 2HNO_3 \rightarrow Ca(NO_3)_2 + 2H_2O$$

Thus, in the zone just ahead of the deterioration front, the pH of the concrete pore solution will be in the range in which nitrifying bacteria populations will undergo growth.

As for sulphuric acid, exposure of concrete to nitric acid leads to mass loss, although this is much more marked when abrasion of the surface is occurring simultaneously. The decalcification of the outer surface of the concrete leads to the formation of an outer layer of amorphous silica gel, followed by an iron-rich zone directly before the unaffected cement which can be seen as a brown layer [101], features which are also seen in the modelled data in Figure 3.16. The thickness of the brown layer is dependent on the concentration of acid, with a higher concentration leading to an increase in thickness. The silica gel layer is prone to shrinkage and undergoes cracking. The rate of deterioration of the cement matrix is strongly dependent on the concentration of acid present (Figure 3.17).

The change in porosity as this occurs would be expected to increase, due to the removal of significant quantities of hydration products. However, experiments on concrete exposed to either nitric acid or a bioreactor environment in which nitrifying bacteria were present indicates only a relatively small change in porosity, with total porosity decreasing in the case of nitric acid exposure (Figure 3.18). It is conceivable that the lack of change in the porosity distribution resulting from exposure to bacteria is the result of pores becoming blocked with biofilm.

Unlike sulphuric acid attack, the relationship between nitric acid resistance and water/cement ratio is the expected one: a higher ratio leads to a microstructure through which the movement of acid is easier, leading to a higher rate of deterioration (Figure 3.19). Additionally, as cement content increases, resistance also increases (Figure 3.20). The results in these two figures are inter-related as a result of the approach taken in designing the concrete mixes: the water/cement ratio was reduced by increasing the

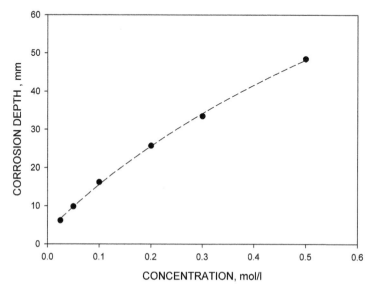

Figure 3.17 Depth of corrosion of Portland cement paste specimens exposed for 200 days in nitric acid solutions maintained at various concentrations [101].

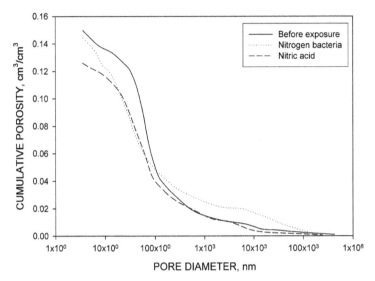

Figure 3.18 Cumulative pore size distributions obtained using mercury intrusion porosimetry for concrete specimens placed in a reactor containing nitrifying bacteria or submerged in a 0.15 mmol/l solution of nitric acid for 14 days [67].

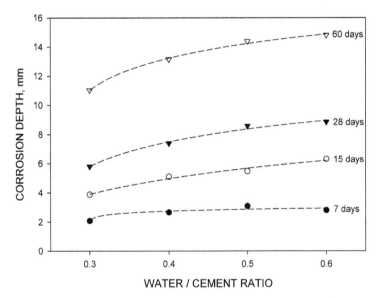

Figure 3.19 Influence of water/cement ratio on deterioration of Portland cement pastes exposed to nitric acid solutions maintained at 0.2 mol/l [102].

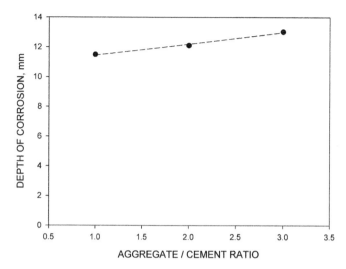

Figure 3.20 Depth of corrosion in 0.5 water/cement ratio Portland cement mortar specimens exposed to a nitric acid solution maintained at 0.2 mol/l after 100 days, w/c = 0.5 mortar [103].

cement content whilst maintaining a constant water content. Nonetheless, the results make sense—a lower water/cement ratio will reduce porosity and pore size leading to a slower rate of ingress, whilst a higher cement content will act to neutralize the acid.

Cement type again plays a role in defining resistance to nitric acid attack. Table 3.8 summarises the findings of experiments in which concrete with different cement types were employed. The outcome of these experiments is less clear cut than for sulphuric acid. Generally, the presence of silica fume has the effect of increasing resistance to nitric acid attack. The presence of siliceous fly ash seems to provide less protection, as do natural pozzolanas.

Calcium aluminate cements are also less susceptible to nitric acid attack in comparison to Portland cement [105]. This is attributed both to the stability of gibbsite to relatively low pH levels, and the enhanced capacity of these cements to neutralize acid [86]. Figure 3.21 shows the result of geochemical modelling of nitric acid attack on a hydrated calcium sulphoaluminate cement. Comparison with the results obtained for Portland cement highlights the persistence of gibbsite, additional quantities of which are precipitated as other calcium aluminate hydration products are dissolved by the acid. This 'band' of gibbsite at the interface between the decalcified and unaffected zones of the concrete is common to all forms of acid attack where acidolysis is the principal mechanism. It has been proposed that precipitation of gibbsite in this way has the effect of blocking porosity and limiting further ingress of acidic species [86].

Table 3.8 Findings of research conducted to evaluate resistance of concrete containing pozzolanic and latent hydraulic materials to nitric acid attack. FA = fly ash; SF = silica fume; NP = natural pozzolana; MK = metakaolin.

Cement		Study				
		[73]	[103]	[104]	[83]	[84]
Slag	Level					
	Acid Conc.					
	Outcome					
FA	Level	7.5–22.5% siliceous fly ash				
	Acid Conc.	0.01%				
	Outcome	Reduction				
SF	Level	7.5–30%	5–30%	0.2%		
	Acid Conc.	All 1%	0.013%	Concentrated and 20%		
	Outcome	Negligible change	Enhancement	Enhancement		
NP	Level				40% 'true pozzolana'	28% trass + 6% calcareous fly ash
	Acid Conc.				2:1 sulphuric/ nitric, pH = 3.5, periodic additions to maintain pH	0.3% to 0.8%
	Outcome				Reduction	Reduction
MK	Level	7.5–30%				
	Acid Conc.	0.01%				
	Outcome	Negligible change				

Whilst water distribution and treatment systems are without question an environment in which damage to concrete from nitrifying bacteria can occur, the effect of the same bacteria on the exterior surfaces of buildings is less certain. In discussing the deterioration of stone masonry, it has been proposed that damage is likely if populations of 10^6 cells/g of material are

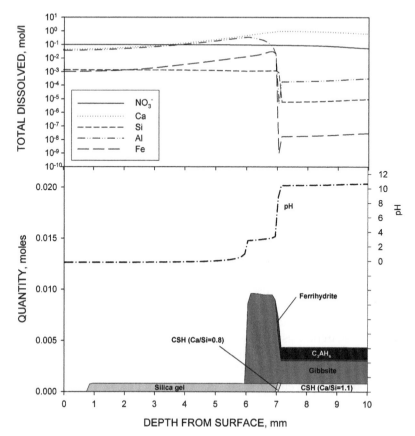

Figure 3.21 Concentrations of dissolved elements (top) and quantities of solid phases obtained from geochemical modelling of nitric acid attack of a typical hydrated calcium sulphoaluminate cement. Model conditions: acid concentration = 0.1 mol/l; volume of acid solution = 4 l; mass of cement 80 g; diffusion coefficient = 5 × 10^{-13} m^2/s.

present [106]. However, this figure comes from a study of deterioration of wall paintings [107], and must be viewed cautiously. Nonetheless, populations of this magnitude can frequently be found on buildings.

A study of stone buildings in Germany found that buildings with significant concentrations of nitrifying bacteria also possessed quantities of nitrate salts at their surface, indicating deterioration had been occurring, albeit potentially over long periods of time [106]. The same researchers also conducted experiments in which concrete blocks were inoculated with nitrifying bacteria and stored under optimal conditions for their growth –30°C, 95% relative humidity and with an excess supply of ammonium nitrate [108]. The concrete blocks had lost more than 3% of their mass after a 12-month period. Deterioration of a much lesser magnitude was observed when identical blocks were exposed to nitric acid at comparable concentrations. However, it should be stressed that this process of biodeterioration was an accelerated one—the provision of more than sufficient substrate in the form of ammonium nitrate does not reflect the situation in reality, where the rate of delivery of ammonia to the building surface will be limited by air pollution levels.

Many nitrifying bacteria can survive and grow using very small quantities of ammonia and nitrite. For instance, *Nitrosospira* from terrestrial locations have half saturation coefficients (K_s, see Section 3.2) as low as 0.1 mg/l NH_3 [108], whilst those of *Nitrobacter* can be as low as 0.06 mg/l NO_2^- [109]. Whilst this means that the bacteria can exist in an environment where very small quantities of ammonia are supplied, it also means that nitrous and nitric acids will be produced in very small quantities, limiting the rate at which damage can be done.

One example of nitrifying bacteria undoubtedly causing damage to a cement-based building surface is that of the deterioration of asbestos-cement roofs on stables [111]. Nitrifying bacteria on the surfaces of such roofs were established as being the cause of deterioration. It should be noted that the circumstances of this case study were favourable to such a process—the presence of horses would have provided a significant source of ammonia, whilst the researchers established that evaporation of moisture within the stable and its subsequent condensation on the underside of the roof acted to leach nitrate salts. Moreover, the elevated temperatures within the stable would probably have encouraged bacterial growth.

3.5.3 Attack by organic acids

When heterotrophic bacteria metabolise organic compounds, the metabolites will often take the form of organic acids. The nature of these acids will depend on the bacteria, environmental conditions and the organic compounds that are being metabolized. There are two main ways in which concrete can potentially be damaged by the bacterial production of organic acids. Firstly, the concrete may be brought into contact with volumes of solutions in which heterotrophic bacteria are breaking down organic material. This is the most likely manner in which deterioration can occur. Secondly, there exists the potential for damage resulting from the growth of bacteria present on the surface of buildings. The former scenario will be explored first.

A number of agricultural activities provide environments where heterotrophic bacteria metabolise organic matter in a volume of liquid. Such processes can either occur circumstantially or intentionally. An example of the intentional formation of organic acids includes the production of silage.

The production of silage—the anaerobic fermentation of grass and other plants for the provision of winter feed for livestock—leads to the formation of organic acids. Traditionally, silage production relied on bacteria already present in the plant matter. Today, however, it is usual for inoculant bacteria to be added to silage to ensure rapid fermentation. The bacteria used fall into two categories: homo- and heterofermentative. Homofermentative bacteria include species such as *Lactobacillus plantarum*, *Enterococcus faecium* and the *Pediococcus* species, and convert the plant sugars exclusively into lactic acid [112]. Heterofermentative bacteria produce a range of metabolites including lactic acid, acetic acid and ethanol. The main heterofermentative species is *Lactobacillus buchneri*.

The mixing of manure with water is an agricultural practice which yields a fluid rich in plant nutrients which can be evenly spread on land with ease. Manure is a substance within which bacterial activity has already occurred, but the storage of liquid manure will usually permit bacteria to continue the process of breaking down the organic compounds, yielding organic acids including propionic, acetic, butyric and isobutyric acid [113].

The production of foodstuffs also generates solutions containing potentially substantial concentrations of organic acids deriving from bacterial activity. Effluents from the dairy industry contain quantities of lactic acid, and often butyric acid [113]. Whey generated from the manufacture of cheese is notably acidic, with pH values around 6, due to the presence of lactic, butyric, propionic and citric acid [114, 115]. The first three of these are caused by bacteria breaking down lactose from milk and are present at concentrations of tens to hundreds of mg/l, whilst citric acid is added to the cheese-making process to curdle the milk and is often present in higher concentrations. Bacteria involved in the formation of the biogenic acids include the *Lactobacillus*, *Streptococcus*, *Pediococcus* and *Leuconostoc* genera.

Thus, the primary organic acids produced by bacteria in applications likely to lead to contact with concrete are acetic, lactic, propionic and butyric acid. The reactions of these acids individually with hydrated Portland cement are discussed below.

Acetic acid

Consulting the data in Chapter 2 regarding the interactions of acetic acid with metal ions in cement, acetic acid forms weak complexes with Ca, Al and Fe(II), and stronger complexes with Fe(III). However, precipitation of salts is typically not an issue, and so the process of deterioration resulting from acetic acid exposure is principally that of acidolysis.

Figure 3.22 shows mass loss from hardened Portland cement pastes exposed to acetic acid solutions, alongside the depth to which deterioration was observed, with both sets of result showing parity with each other, and also fairly substantial damage in a relatively short period. Whilst being acidolysis-based, and hence producing results comparable to nitric acid

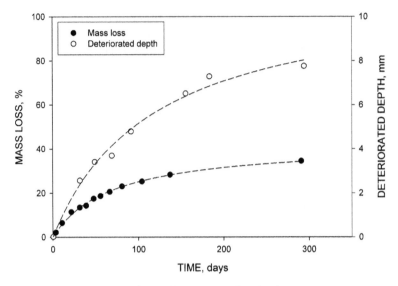

Figure 3.22 Mass loss and depth of deterioration of Portland cement paste specimens exposed to acetic acid [116]. Acid concentration = 0.28 mol/l, with pH adjusted to 4.0 using sodium hydroxide.

attack, deterioration will typically occur at a slower rate with acetic acid, since it is a weaker acid.

Since acidolysis is the main mode of attack, deterioration takes the form of decalcification at the surface, leading to a thick layer of what is essentially silica gel being the only remaining material at the outside of the cement paste. Figure 3.23 shows micro-CT scans through various cement paste specimens exposed to acetic acid solutions. The dark outer layer is the silica gel layer. In the case of the Portland cement specimen, there is a second layer between the wholly-decalcified material and the unaffected cement, where partial decalcification has occurred, which is not present in the case of the other cements in this figure, due to their lower calcium content. It should be noted that the silica gel layer has cracked substantially in all cases. This is not the direct result of acid attack, but of shrinkage during drying. However, the magnitude of the cracking shown in these images clearly illustrates the notable susceptibility of the layer to shrinkage.

Taking the diameter of the wholly unaffected cement as a measure of resistance, the figure indicates that there is some enhancement in resistance to acetic acid attack through the use of fly ash. This has also been observed with concrete [117], with enhanced performance with silica fume reported frequently, also (Table 3.9). Calcium aluminate cements a sulfoaluminate cement—CSA—in the case of Figure 3.23 typically provide enhanced resistance, for the same reasons as for nitric acid.

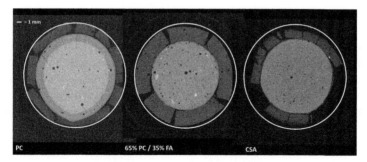

Figure 3.23 micro-CT scans from cement paste specimens exposed to a 0.1 M solution of acetic acid for 90 days [118].

Table 3.9 Findings of research conducted to evaluate resistance of concrete containing pozzolanic and latent hydraulic materials to acetic acid attack. FA = fly ash; SF = silica fume; NP = natural pozzolana; MK = metakaolin.

Cement		Study				
		[117]	[103]	[73]	[80]	[81]
FA	Level	5%				
	Acid conc.	1.7%				
	Outcome	Enhancement				
SF	Level	10%	5–30%	7.5–30%	78.5–30%	15%
	Acid conc.	1.7%	14%	5%	5%	5%
	Outcome	Enhancement	Enhancement	Reduction	Enhancement	Enhancement
MK	Level			7.5–30%		
	Acid conc.			5%		
	Outcome			Negligible change/ Reduction		

Lactic acid

Lactic acid does not form insoluble salts with calcium, aluminium or iron, and forms weak complexes with these ions (although the aluminium complexes are somewhat stronger than those with other ions). For this reason, the deterioration of concrete in contact with lactic acid is almost purely a process of acidolysis.

An example of mass loss characteristics from lactic acid attack are shown in Figure 3.24, whilst micro-CT scans from cement paste specimens made from PC, PC/FA and CSA cements are shown in Figure 3.25. It should be noted that the specimens in this figure show less deterioration than those exposed to acetic acid. From a theoretical perspective, this seems improbable, since lactic acid has a lower acid dissociation constant (see Chapter 2) and is, thus, the stronger of the two. A study examining the composition of organic acid solutions brought into contact with hardened Portland cement paste [120] has found that, despite the lower pH of the lactic acid solution, the concentration of calcium is lower than that of acetic acid (as well as butyric, isobutyric and propionic acids).

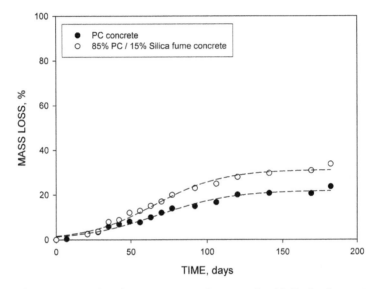

Figure 3.24 Mass loss from concrete specimens made with Portland cement and a Portland cement/silica fume blend exposed to lactic acid [81]. Specimens were scrubbed periodically with a wire brush.

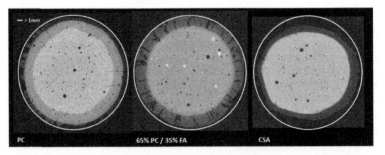

Figure 3.25 micro-CT scans showing cross-sections through cement paste specimens exposed to a 0.1 M solution of lactic acid for 90 days [119].

The reason for this is almost certainly related to the fact that the lactate ion forms complexes with Al, leading to an increase in the concentration of this element in solution. Higher Si concentrations were measured compared to the other acid, which is at least partly the result of the lower pH obtained with lactic acid. However, neither of these effects directly explain the lower Ca concentrations. It is possible that, since CSH gel will normally contain a quantity of aluminium, the removal of this element from the gel as a result of complexation means that calcium takes its place.

Both figures show that the use of pozzolanic material and CSA enhances resistance to lactic acid attack. Slag also gives improved resistance, as will be discussed later in this section.

Propionic acid

Very little research has been conducted into the interactions of propionic acid with cement and concrete. However, experiments using of a number of the common organic acids formed by bacteria have been conducted to examined the pH and dissolved species detected after exposure of crushed PC paste specimens to acid solutions of the same molar concentration [120]. It was found that there was very little difference between the behaviour of propionic acid compared with either acetic, butyric or isobutyric acid. This is not surprising, since all of these acids are structurally similar and have very similar acid dissociation constants (see Chapter 2).

Butyric acid

As previously stated, butyric acid's chemical similarity to acetic acid means that the nature of its interaction with hardened cement is comparable. This can be seen in Figure 3.26, which shows very similar cross-sections through cement paste specimens as those seen in Figure 3.23.

Organic acid mixtures

Whilst it is useful to understand the effects of these organic acids on concrete individually, it must be remembered that they commonly occur as mixtures. For this reason, a number of studies have been carried out using such mixtures. Because of the similar chemical behaviour of acetic, butyric, isobutyric and propionic acid [120], the approach taken has been to use acid solutions comprising acetic and lactic acid. Cement types studied have included PC, PC/GGBS [121], PC/fly ash and PC/silica fume cements [122]. A comparison of mass loss from PC and PC/GGBS mixes are shown in Figure 3.27, with clearly enhanced performance where GGBS is used. Also illustrated in this figure is the influence of cement content, which appears

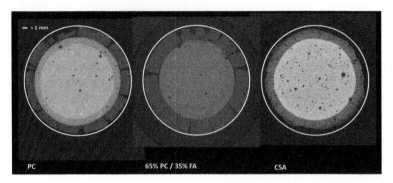

Figure 3.26 micro-CT scans showing cross-sections through cement paste specimens exposed to a 0.1 M solution of butyric acid for 90 days [118].

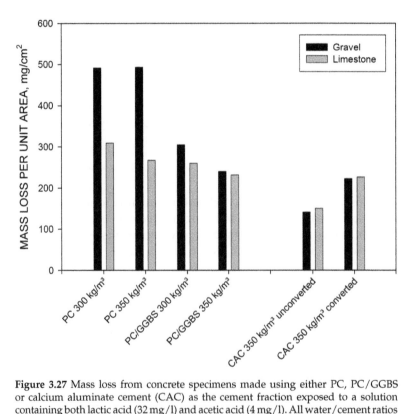

Figure 3.27 Mass loss from concrete specimens made using either PC, PC/GGBS or calcium aluminate cement (CAC) as the cement fraction exposed to a solution containing both lactic acid (32 mg/l) and acetic acid (4 mg/l). All water/cement ratios are 0.45, except the CAC mixes (w/c = 0.40). PC/GGBS proportions unstated, but by definition must be 36–95% by mass [121].

to have little effect in the case of the PC-only mixes, but which produces lower levels of deterioration for lower cement content where GGBS is present.

Figure 3.28 compares performance in volume reduction terms of PC, PC/FA and PC/SF mixes of the same water/cement ratio. Silica fume is more effective than fly ash, although both give higher resistance to deterioration compared to PC [122]. Generally, performance increases in the sequence PC–PC/FA–PC/SF–PC/GGBS [123].

Calcium aluminate cement also seemingly provides greater resistance to attack (Figure 3.27), although it should be noted that the CAC mixes in this figure had lower water/cement ratios, which is also likely to play a part. Conversion of the cement (the result of carbonation, leading usually to a loss in strength) reduces resistance slightly, but the converted material still performs better than the PC-only mixes [121].

The influence of aggregate type is also shown in Figure 3.27. The use of limestone aggregate provides greater resistance when compared to siliceous gravel for the PC and PC/GGBS mixes. This is the result of the calcium carbonate present in limestone increasing the neutralising capacity of the concrete. This effect is not evident in the case of calcium aluminate cement.

In a similar study, a solution mimicking the composition of pig manure was used [124]. This comprised 0.21 mol/l of acetic acid, 0.04 mol/l propionic acid, 0.02 mol/l butyric acid, 0.01 mol/l isobutyric acid

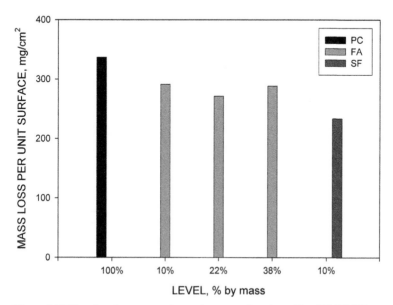

Figure 3.28 Mass loss from concrete specimens made using either PC, PC/FA or PC/SF as the cement fraction exposed to a solution containing both lactic acid (30 mg/l) and acetic acid (30 mg/l). All water/cement ratios are around 0.40 [122].

and 0.003 mol/l of valeric acid. The solution was regularly refreshed to maintain these concentrations. Figure 3.29 shows how the development of deteriorated depth settles into what approximates to a linear relationship with respect to time under these conditions. The pH of these solutions was adjusted to 4 and 6 using sodium hydroxide, and it is evident that pH plays an important role, with a high pH solution yielding less damage. GGBS was found to improve performance. 10% Silica fume also enhanced resistance, albeit to a slightly lesser extent [125].

Heterotrophic bacterial attack in the laboratory

An interesting example of bacterial deterioration observed in the laboratory involved bacterial samples taken from a deteriorating sewer [126]. The sewer was suspected of being attacked under the action of sulphate reducing and sulphur oxidising bacteria. Bacteria were isolated from samples taken from surfaces in a sewer system and transferred to bioreactors in which mortar specimens were held in previously sterilized growth culture solution. Initially, sulphate reducing bacteria were introduced, followed by sulphur oxidising organisms. The concrete underwent deterioration, but analysis of the deteriorated material found an absence of gypsum and ettringite. However analysis of the growth culture medium found high concentrations of acetic, propionic and, by inference, carbonic acid. Examination of the isolated bacteria indicated that the sulphate reducing bacteria population

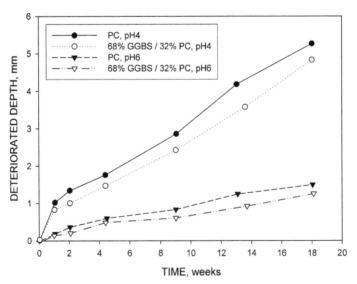

Figure 3.29 Deteriorated depth versus time for PC and PC/GGBS pastes in a mixed acetic/propionic/butyric/isobutyric/valeric acid solution [124].

consisted of two different species, one of which could not be identified and either a species of *Flavobacterium* or *Xanthomonas*. The sulphur oxidizing bacteria were *Thiobacillus intermedius*. It was concluded that the concrete in the bioreactors was deteriorating as a result of attack from organic acids deriving from aerobic bacteria (and, thus, presumably *Flavobacterium* or *Xanthomonas*). It should be stressed that this does not mean that the sewer pipes themselves were undergoing this form of deterioration, simply that the samples taken and the conditions of the bioreactor favoured the development of aerobic heterotrophic organisms.

Studies into the effect of heterotrophic bacteria on reinforced concrete have observed considerable deterioration of properties and accelerated de-passivation of the steel [127, 128]. Concrete specimens submerged in a growth medium solution in which heterotrophic bacteria were present were found to display lower strengths than specimens submerged in sterilised water from the same source [127]. The disparity between the strength was as high as 36%, with the greater difference in strength corresponding to the weaker concrete mixes, although the researchers do not state the period of time over which this occurred, or whether the differences correspond to losses in strength or arrested strength development. Electrochemical measurements made on the steel reinforcement bars embedded in other specimens found that the corrosion potential in the presence of bacteria increased at early ages and, in most cases, remained high, whereas the sterile specimen potentials remained low. However, potentiodynamic polarisation measurements indicated passivation of the steel affected by bacteria, and it was proposed that this was the result of biofilm formation at the steel surface.

A similar study did not use a nutrient solution, but instead introduced sodium citrate into the concrete as an admixture at the mixing stage [128]. The citrate ion can be used as a nutrient by heterotrophic bacteria. Sodium citrate is used as a corrosion inhibitor, but not normally in concrete, since it will lead to the formation of calcium citrate leading to expansion and cracking. Consequently, the 28 day strengths of the concrete mixes were reduced in direct proportion to the admixture dosage. The concrete specimens were submerged in volumes of water removed from a pond —to ensure the presence of heterotrophic bacteria—for a period of 90 days. The strength of all specimens increased over this period, although strength gain was proportionally less in the specimens containing higher quantities of citrate. Corrosion potential measurements indicated a rise in corrosion potential with citrate dosage, and gravimetric measurements on the reinforcement indicated a higher rate of corrosion above sodium citrate dosages of 1.0%.

Whilst both these last two studies suggest that the presence of heterotrophic bacteria have the potential to cause significant damage to concrete and its reinforcement, features of the experimental design and interpretation of results in both cases means that this aspect still needs further investigation.

3.5.4 Biofilm formation as a form of deterioration

The growth of biofilms on building surfaces also potentially has aesthetic implications for concrete surfaces. The main bacterial contribution towards discolouration of construction material surfaces comes from cyanobacteria.

Often the most significant proportion of biomass on a building surface will comprise cyanobacteria. A study of building surfaces in both Europe and Latin America found cyanobacteria as the dominant microbe at many sites [129]. The families of cyanobacteria identified were *Scytonemataceae* (the most common), *Microchaetaceae*, *Rivulariaceae*, *Oscillatoriales* and *Nostocaceae*. Cyanobacteria have also been identified as the main component of stains on concrete walls in Toulouse in France. In this case, the genera of *Gloeocapsa-Chroococcus* (family: *Chroococcaceae*), *Microspora* (family: *Microsporaceae*) and *Trentepohlia-Gloeocapsa* (Family: *Microcystaceae*) were identified [32]. A programme of sampling from building façades around France identified the cyanobacteria genera *Aphanocapsa, Aphanothece, Calothrix, Chroococcus, Cyanosarcina, Gloeocapsa, Gloeothece, Leptolyngbya, Microcoleus, Nostoc, Phormidium* and *Scytonema* [130]. Analysis of DNA from concrete surfaces from sites in Georgia, USA has also identified cyanobacteria as major occupiers of these surfaces [30].

Cyanobacteria, as the name suggests, often have a blue-green pigmentation, which is the result of the presence of phycobilisomes— regions in the inner membranes of the bacteria which capture energy from light. These contain phycobiliproteins and carotenoids, which give the bacteria their colour [3]. Colours are not limited to blue-green: carotenoids and a group of phychobiliproteins known as phycoerythrins impart a reddish-brown colour.

The presence of substantial quantities of cyanobacteria on the surface of a building material is therefore likely to be visible, potentially in a manner which is undesirable. Cyanobacteria from the genus *Gloeocapsa-Chroococcus, Microspora* and *Trentepohlia-Gloeocapsa* have been observed to produce black, green and red stains respectively [32].

Both environmental and material characteristics influence colonisation by cyanobacteria. The previously mentioned study of European and Latin American cyanobacteria found that the dominance of cyanobacteria was more pronounced for concrete, mortar and brick surfaces in the Latin American locations, and it was proposed that this was the result of these organisms ability to resist dehydration and higher temperatures. Nonetheless, moisture is an important factor and the high humidity of the Latin American locations was thought to favour growth of cyanobacteria. This is also reflected in the tendency of cyanobacteria to grow preferentially on the façades of buildings which are facing the dominant wind direction (which are consequently most likely to be the dampest side of a building) [130]. The sides of building receiving less sunlight during the day will also tend to retain moisture longer, promoting growth of cyanobacteria.

The material characteristic which has most influence over cyanobacterial colonisation is the porosity of concrete, with higher levels of porosity (corresponding to higher water/cement ratios) encouraging growth [30, 31, 32]. Surface roughness, as discussed earlier in this chapter, appears to have less of an influence [30, 32]. Portland cement composition and the use of pozzolanic and latent hydraulic materials have little influence over colonization rates [30].

It should be stressed that biofilms are frequently not exclusively composed of bacterial communities, and can contain a range of different organisms. For this reason, biofilms will be revisited in subsequent chapters.

3.5.5 Aggressive CO_2

As seen earlier, heterotrophic bacteria will ultimately oxidise organic compounds to CO_2, potentially creating an environment in which aggressive CO_2 is present. This is certainly possible in the lower level of water (the *hypolimnion*) in lakes. In such environments, the decomposition of organic matter leads to the dissolution of calcium carbonate in the sediment by aggressive CO_2, leading to the sediment becoming enriched in elements such as iron [131].

Attack of concrete used in the gate structure of a shipway in Virginia, USA is thought to be the result of combined sulphate attack and attack by aggressive CO_2 [132]. Samples taken from the water indicated high levels of aggressive CO_2 (up to 57 mg/l) and a pH as low as 6.9. Reference to Chapter 2 indicates that such a pH is sufficient to significantly increase the solubility of calcium ions. Whether the high concentrations of CO_2 are ascribable to bacterial activity was not explored, but given that conditions in such waterways may well be similar to those in natural lakes, it is reasonable to suppose that this is a possibility.

In the discussion of organic acid formation by heterotrophic bacteria, the study of bacteria isolated from a deteriorating sewer produced not only organic acids, but also carbonic acid [126]. This was established as a result of the formation of large quantities of calcium carbonate at the surface of mortar specimens in contact with the bacteria. Whilst the concentration of carbonic acid was not established, the low pH obtained (as low as 6.2) suggested that this was partly the result of the presence of carbonic acid.

The influence of carbonic acid on the loss of mass from concrete is shown in Figure 3.30. The figure shows the effect of periodic brushing of the specimens in comparison to specimens which experienced no abrasion, with mass loss from the abraded specimens being considerably higher. The results

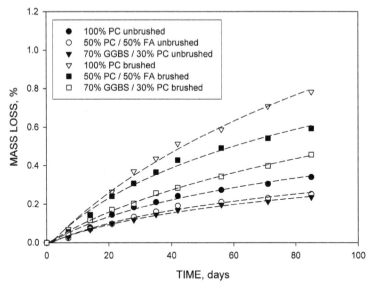

Figure 3.30 Mass loss from concrete specimens exposed to a carbonic acid solution with a pH of 4.1 [133].

from three different concrete mixes are shown—PC, and 50% PC/50% FA and 70% GGBS/30% PC. The GGBS and FA mixes show greater resistance to attack, with the GGBS mixes performing particularly well when abrasive conditions are effective. It should be noted, however that rates of mass loss are considerably smaller than for other acids discussed previously.

3.5.6 Long-term processes

Most of the issues relating to bacterial deterioration of concrete that have been discussed in this chapter concern timescales pertinent to the built environment—years to hundreds of years. However, in some applications, persistence of integrity of cement-based materials over much longer periods are necessary. The most obvious of these is in nuclear waste repositories where low and intermediate level wastes may be encapsulated in a cementitious matrix within a metal container for storage for periods of many thousands of years. Moreover, the repository is normally backfilled with cementitious material after the waste is deposited to create a high pH environment that limits the likelihood of leaching of radionuclides.

Engineers involved in planning such activities clearly need to consider potential damage to such 'waste packets' by bacterial activity, since even small-scale bacterial activity over these timescales presents the potential for significant deterioration.

From a nutrient perspective, low and intermediate level nuclear waste will frequently contain organic carbon in the form of paper, wood and cotton [134]. Whilst the compounds in these materials are not directly useable by bacteria, they are able to produce enzymes which break the compounds down to smaller molecules [135]. Additionally, the high pH conditions have the potential to lead to alkaline hydrolysis of cellulose to form compounds that can be metabolized by bacteria [134].

Another organic material that may be present in a nuclear waste repository is bitumen, which is used in some cases as the encapsulation medium. Leaching of this material has been found to release organic compounds including alcohols, carbonyl compounds, glycols, aromatic compounds, nitrogen compounds and carboxylic acids, which could potentially be used by bacteria [136]. Moreover, exposure to radioactivity can yield a greater quantity of these compounds in leachate [137]. Where the leachant has a composition comparable to water which has been in contact with Portland cement, levels of leaching are also higher. Bituminized waste packages will also frequently contain a source of nitrogen in the form of sodium nitrate, which is highly soluble [138].

A study has examined the effects of simulated leachate from a bituminized waste package coming into contact with a CEM V cement (containing a combination of Portland cement, slag and fly ash) [139]. It found that where the acids formed included oxalic acid, this was not available for bacterial use, since it was rapidly rendered insoluble in contact with cement (see Chapter 2). However, acetic acid and nitrate remained available to bacteria.

For heterotrophic bacteria, the often relatively abundant presence of carbon means that the limiting substrates in repositories are likely to be inorganic substances used as sources of oxygen, such as sulphate and nitrate.

Since repositories are designed to wholly isolate their contents from the external environment, they will be sealed to the external atmosphere. Thus, oxygen will gradually be used up by aerobic bacteria, to be replaced with CO_2. Once oxygen is exhausted, bacteria will begin to reduce any available nitrate to obtain oxygen, followed by sulphate when this is also used up [140]. Analysis of gases in sealed containers in which such processes are occurring indicates an accumulation of methane once sulphate is exhausted, indicating a shift towards anaerobic respiration [134]. The resulting reducing atmosphere is considered to be favourable for repository safety, since this not only limits the extent to which corrosion of the steel containers can occur, but also tends to significantly reduce the solubility of many radionuclides [141].

Models of the growth of bacteria in repository environments is able to reproduce the features described above [134, 140], with one model predicting bacterial activity for around 400 years after the repository is completed [134]. On the basis that the majority of bacterial activity involves heterotrophic bacteria, the formation of organic acids presents the main threat from bacteria in repositories.

Experiments where fluid was percolated through columns of simulated intermediate level waste and crushed cement paste have found that bacterial activity led to concentrations of between 0.07 and 0.7 mmol/l of acetic, propionic and butyric acid [134]. Whilst these concentrations are relatively low, it must be remembered that this does not mean that damage will not be done. Taking the example of mixed organic acid solutions shown in Figure 3.29, a deterioration depth of around 5.25 mm is obtained after 18 weeks. If a linear relationship between the depth of deterioration and time is assumed, and the rate of movement of this front is assumed to be in direct proportion to the concentration of acid, using a concentration of 0.7 mmol gives a depth of deterioration of around 15 mm after 400 years. This represents a worst-case scenario, since it would be expected that bacterial activity would fall off as sources of carbon, nitrate and sulphate become scarcer. Nonetheless, it is clear that the use of cementitious backfill is a prudent strategy to act as an additional protective measure.

Another area in which long-term deterioration by bacterial activity is a potential problem is in the decommissioning of oilwells. The production of hydrogen sulphide by sulphate reducing bacteria is frequently encountered process in oilwells. The resulting 'sour' conditions are of concern because they are associated with the corrosion of steel casings, etc. The corrosion process has been proposed to be the result of the reaction [142]:

$$(x{-}1)Fe + S_{y-1} {\cdot} S^{2-} + 2H^+ \rightarrow (x{-}1)FeS + H_2S + S_{y-x}$$

This reaction undoubtedly occurs because layers of iron sulphide can be detected at steel surfaces. However, an alternative theory is that, for corrosion to occur at the rates that can be observed, it is more likely that elemental sulphur forms sulphuric acid. This can occur to some extent through hydrolysis [143]:

$$S_8 + 8H_2O \rightarrow 6H_2S + 2H_2SO_4$$

However, research into the effect of adding nitrates to sour wells to limit sulphide production has found that, despite limiting sulphide levels, the rate of corrosion of steel increases considerably with nitrate addition [144]. This suggests that bacterial activity is the cause of the corrosion. Specifically, it implies that sulphur oxidising bacteria are utilising nitrate as a source of oxygen, forming sulphuric acid which is the source of the steel corrosion. Sulphur oxidising bacteria are undoubtedly present in such environments [145].

The significance from a concrete (or more accurately, cement) perspective relates to the plugging of oilwells once they have ceased to be viably productive. Such plugging activities require sealing the well for geological periods of time, as is the case for nuclear waste storage. The material of choice for the plug (or 'barrier') is currently cement, although other materials can potentially be used [146]. Whilst the process of cement plug deterioration by sulfuric acid attack is one that is a potentially real one, it should be put into perspective—multiple barriers with combined lengths of 50 to 100 m are typical. Additionally, a source of oxygen is necessary to produce sulphuric acid. This will normally derive from nitrate, but since concentrations of nitrate in such environments are typically limited, it is likely that sulphuric acid formation would only occur for relatively short periods of time. Nonetheless, in the UK, evaluation of barrier materials requires testing under conditions which mimic downhole conditions to establish rates of deterioration, to allow prediction of performance [146].

The organic acids found in waters associated with oil formations are characteristic of those produced by heterotrophic bacteria. However, their origins are most probably related to the exposure of oil hydrocarbons to elevated temperatures, rather than any biological degradation [147].

3.6 Limiting Bacterial Deterioration

When considering options available for protecting concrete from bacterial biodeterioration, some measures are specific to the type of bacteria involved. This is particularly true of attack by sulphur bacteria where behaviour of concrete is, in some respects, different than for other forms of bacterial attack. Moroeover, the processes which lead to attack of concrete by sulphur bacteria may be addressed not only through engineering of the concrete, but through manipulation of the environment in which the bacteria live. For this reason, approaches to enhancing the durability of concrete are discussed below, firstly in terms of those which are apply solely to attack by sulphur bacteria, and, secondly, in terms of those which are generic.

3.6.1 Sulphur bacteria

Two different approaches are available with regards to limiting the damage from sulphur bacteria. Firstly, the environment in which sulphur bacteria are likely to establish themselves can potentially be engineered such that the concentrations of substances which play a part in the deterioration process are limited. Secondly, concrete itself can be designed and possibly treated in such a manner that it possesses enhanced resistance.

Environmental control

The manner in which sewers are designed and operated can be used to limit damage from sulphur bacteria. The removal of H_2S before it comes into contact with the sewer walls is clearly a desirable condition, and this can be achieved through ventilation of the air space in the sewage pipe. Ventilation is normally achieved through natural ventilation processes which utilise wind, air pressure differences within the system, and air movement resulting from frictional drag from moving wastewater or suction resulting from a fall in wastewater level [148]. Where issues of H_2S accumulation are serious, mechanical ventilation may also be employed. Ventilation has additional beneficial side effects. Firstly it maintains high concentrations of oxygen in the sewer atmosphere, which will be seen later to be beneficial. Secondly, the movement of air will often have a drying effect on sewer walls, leading to an environment low in moisture, thus creating conditions unfavourable for colonisation by sulphur oxidising bacteria.

Since accumulation of H_2S in the air-space above the waste water is a key aspect of the attack process, one means of preventing this happening, in theory, is to simply remove the airspace by running the sewer full [89]. This essentially entails the use of force mains, siphons and surcharged sewers. However, the alternative viewpoint is that this will create anaerobic conditions which promote sulphide formation by sulphate reducing bacteria, and this approach is therefore not wholly ideal [149].

The alternative is to ensure that the conditions within the sewer are sufficiently aerobic to limit the growth of sulphate reducing bacteria, and to increase the redox potential, allowing for oxidation of sulphide to sulphate. This is normally done by ensuring that the rate of flow is sufficiently high. The rate of increase in the concentration of sulphide in a pipe which is running partially full can be described using the empirical equation [150]:

$$\frac{d[S]}{dt} = \frac{M[EBOD]}{r} - \frac{N[S](su)^{3/8}}{d_m}$$

where
[S]	=	sulphide concentration (g/l);
M,N	=	empirical coefficients;
[EBOD]	=	effective biochemical oxygen demand (g/l);
r	=	the radius of the pipe (m);
u	=	stream velocity (m/s);
s	=	the slope of the pipe, expressed as a fraction; and
d_m	=	hydraulic depth (m).

Biochemical oxygen demand (BOD) is the amount of oxygen needed by aerobic microbes to break down the organic matter in a sample of water, expressed as a mass of oxygen per volume of water. This is also dependent on temperature, and so effective BOD (EBOD) includes a means of describing the effect of temperature:

$$[EBOD] = [BOD](1.07)^{T-20}$$

where T is temperature expressed as °C. A higher $[EBOD]$ will increase the rate at which oxygen is exhausted, thus creating anaerobic conditions appropriate for sulphate reducing bacteria.

Thus, from the rate equation it is evident that increasing the velocity of the stream and/or increasing the slope of the pipe reduces the rate of development. This is believed to be for two reasons. Firstly, an increased rate of flow will increase the rate at which re-aeration occurs, thus providing adequate oxygen to satisfy the BOD and prevent aerobic conditions establishing. Secondly, faster flow will tend to act to scour biofilms and solids at the bottom of the pipe, limiting their development as a habitat for sulphate reducing bacteria. However, some doubts have been expressed with regards to how important this process is [148].

Where turbulence is created, re-aeration rates are further enhanced, although it must be noted that where turbulence occurs and sulphide is already present, this tends promote the release of H_2S as gas [149].

Dissolved oxygen can be increased further by injecting air or pure oxygen into the water, usually in features within a sewer system such as pressurized mains pipes ('force' or 'rising' mains) or the sumps of gravity flow sewers (wet wells) [152]. The biofilm (or slime layer) on the sewer wall in which sulphate reducing bacteria reside will act as a barrier to oxygen. However, increasing the dissolved oxygen concentration to above 0.5 mg/l usually has the effect of increasing the depth of penetration sufficiently to limit bacterial growth and allow sulphide oxidation [149]. The effect of air injection is shown in Figure 3.31.

Given that this is the result of a process of oxygen diffusion through the biofilm, increasing the concentration of dissolved oxygen in the water will

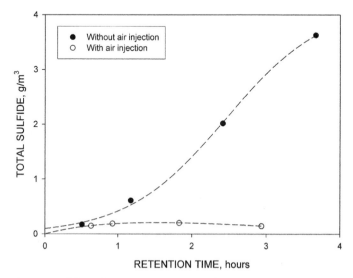

Figure 3.31 Effect of air injection on sulphide concentrations in a sewer at different points long its length (expressed in terms of retention time wastewater has spent in system) [152].

accelerate the reaction, since it will create a greater concentration gradient. Thus, there is some justification in using pure oxygen in comparison to air, although given the considerable quantity of energy required to purify oxygen from air, the environmental credentials of this are likely to be poor.

The rate at which sulphide is produced by bacteria can be reduced by the addition of soluble nitrate salts (typically sodium nitrate) to wastewater. A number of theories exist as to why this is effective. When nitrates are added to wastewater, development of populations of chemolithotrophic bacteria which oxidise sulphide whilst reducing nitrate is observed [153]. However, the alternative explanation is that the addition of nitrate promotes the growth of heterotrophic nitrate reducing bacteria which, whilst not having a direct influence on sulphide production, compete with sulphate reducing bacteria to limit its formation [154]. Regardless of which process occurs (and it is conceivable that both occur simultaneously) nitrate is observed to be reduced in preference to sulfate in oxygen-poor sewer environments [155].

Various oxidant compounds can potentially be introduced into the sewer systems to oxidise hydrogen sulphide to sulphur. Hydrogen peroxide (H_2O_2) is one of these compounds. It undergoes the reaction:

$$H_2O_2 + H_2S \rightarrow S + 2H_2O$$

From the equation, the minimum molar ratio of H_2O_2 to H_2S is required to be 1, but may need to be as high as 5 to ensure thorough removal [149]. Chlorine in the form of chlorine gas or a solution of a hypochlorite compound can be used in a similar manner:

$$Cl_2 + H_2S \rightarrow S + 2HCl$$

Despite a theoretical need for a molar ratio of Cl_2 to H_2S of 1.0, ratios between 10 and 15 are required, which can make the cost excessive. Chlorine is a hazardous gas, and the risk of leakage into the wider environment is a valid concern which limits its use. Chlorine and hypochlorite also have a biocidal effect. However, their ability to kill bacteria is limited by the effectiveness of biofilms in protecting resident bacteria, again requiring high concentrations to be effective. A number of instances of bacterial resistance deriving from the presence of biofilms have been observed, albeit not for sulphate reducing bacteria [151, 156, 157].

Another oxidant, potassium permanganate, has been identified as being suitable. Under acidic conditions it takes part in the following reaction:

$$2KMnO_4 + 3H_2S \rightarrow 3S + 2H_2O + 2KOH + 2MnO_2$$

Where conditions are alkaline, potassium sulphate (K_2SO_4) is also formed. Potassium permanganate is relatively expensive and must be used at $KMnO_4$ to H_2S ratios of 6–7, making it a less common practice [149].

In any sewer, a proportion of sulphide will be present in insoluble form as metal sulphide salts. Metals which form insoluble sulphides include iron

(II), manganese, zinc, copper and nickel. One possible means of reducing hydrogen sulphide formation is to add additional of soluble metal salts to wastewater which will precipitate sulfide salts. The most appropriate candidates are iron (II) salts, since they are relatively cheap and are likely to have lesser environmental implications, due to their lower toxicity in comparison to the other suitable metals.

The addition to wastewater leads to a reaction of the form:

$$Fe^{2+} + HS^- \rightarrow FeS + H+$$

However, addition of Fe (III) salts is as effective, if not more so (Figure 3.32), since in wastewater Fe (III) will be reduced to Fe (II):

$$2Fe^{3+} + HS^- \rightarrow 2Fe^{2+} + S^0 + H^+$$

Suitable soluble iron (II) compounds are iron (II) chloride ($FeCl_2$), iron (II) nitrate ($Fe(NO_3)$) and iron (II) sulphate ($FeSO_4$). Suitable iron (III) compounds are the ferric equivalents: $FeCl_3$, $Fe(NO_3)_3$ and $Fe_2(SO_4)_3$.

It has already been shown that the formation of H_2S requires acidic conditions, and so raising the pH will limit its formation. A pH of greater than 9 is required, since this will ensure that the vast majority of dissolved sulphide is present as the dissociated HS^- rather than H_2S (Table 3.2). This can be achieved through the addition of sodium hydroxide (NaOH). Continuous dosing is prohibitively expensive [159], but an alternative approach is to add large doses to achieve a pH of 12.5–13.0 for periods of

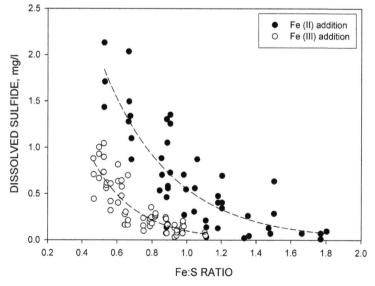

Figure 3.32 Effect of the addition of Fe (II) and Fe (III) salts on dissolved sulphide concentrations in wastewater samples [158].

less than an hour [160]. These conditions will reduce H₂S formation, but also have the effect of severely reducing the population of sulphate reducing bacteria in a sewer system for a number of days, after which the process must be repeated (see Figure 3.33). One problem with this approach is that the high pH wastewater resulting from this 'shock' dosing cannot undergo the microbial digestion treatments that sewage will normally undergo downstream, and so must be diverted and dealt with by alternative means.

Magnesium hydroxide (Mg(OH)₂) has been used in a similar manner to sodium hydroxide [161]. It has the benefit of being 'self-buffering'—it maintains pH at around 9.5 to 10.0 despite perturbations in the form of additions of acidic or alkaline substances. This is because magnesium hydroxide precipitates as a solid at a pH above around 10.0. Thus, if the pH begins to rise, magnesium hydroxide will precipitate, removing OH⁻ ions from solution and reducing the pH. As the pH falls below 10.0, solid magnesium hydroxide dissolves again, maintaining the pH above 9.5. This characteristic is beneficial in that it prolongs the high pH conditions, thus extending the period over which dosing is effective.

Cleaning of sewer pipes has the potential to remove both the biofilms and deposits of solids in which sulphate reducing bacteria reside, along with sulphur oxidizing bacteria on pipe surfaces above the waterline. A range of different techniques can be used, which are selected on the basis of the size of the sewer, the extent to which deposits have established themselves, and the types of deposits present [163].

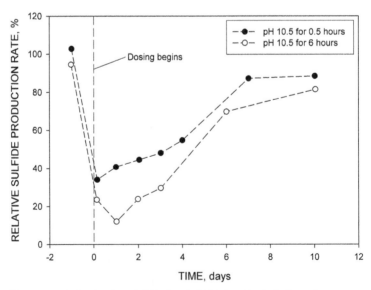

Figure 3.33 Reduction in sulphide levels following shock dosing of a pressure main sewage system with sodium hydroxide for two different durations [162].

Hydraulic cleaning techniques employ the movement of water, and include flushing, jetting and balling. Flushing simply involves increasing the rate of flow through pipes for a short period of time. In the case of jetting, a high power spray of water is directed at the pipe sides. Balling involves pulling a ball that is slightly smaller than the pipe diameter through the sewer whilst water is made to flow past it. This has the effect of causing the ball to spin round within the confines of the pipe, scraping material from the surface in the process.

Mechanical cleaning techniques include power rodding and winching. Power rodding employs a drive unit to push a long flexible rod through the pipe. At the end of the rod is a blade configuration which rotates to loosen debris. Winching employs a bucket, which consists of a cylinder closed at one end, with two opposing jaws at the other end which can be opened up such that when pulled through a pipe they dislodge debris, which are then collected in the bucket.

Cleaning is not only conducted to reduce bacterial attack, but is a more general maintenance activity employed as a means of clearing blockages and as a measure to prevent such blockages occurring. However, it has been found to be effective in reducing the rate of sulphide oxidation. The regularity with which cleaning must be conducted to be effective in controlling bacterial deterioration will depend on the conditions within a sewer. However, one study found that sulphide oxidation rates were restored after only 10–20 days [164].

In the case of H_2S production in oil wells, other chemicals can be introduced into sewer systems which are specifically used for their ability to kill bacteria—biocides. Two of the most commonly used biocides are diamines, most commonly cocodiamine, and glutaraldehyde [165]. However, bacteria are well-known for their ability to rapidly adapt to hostile conditions, and the development of resistance to diamine biocides has been observed [166].

Molybate compounds (such as ammonium heptamolybdate, $(NH_4)_6Mo_7O_{24} \cdot 4H_2O$) and nitrite compounds (such as sodium nitrite, $NaNO_2$) act as metabolic suppressors to sulphate reducing bacteria [165, 167]. Research has identified that their use in combination is potentially the most economic way of employing them (Figure 3.34) [165]. This approach has also been successfully applied to reduce hydrogen sulphide emissions from manure slurries arising from livestock production [168].

Research has also examined the possibility of using genetically-engineered bacteria to produce biocidal substances to limit the growth of sulphate reducing bacteria [169]. The bacteria used for this purpose were strains of *Bacillus subtilis* which were genetically engineered to secrete the peptide antimicrobials indolicidin and bactenecin. When introduced into cultures containing the sulphate reducing bacteria *Desulfovibrio vulgaris* this population was reduced by 83%.

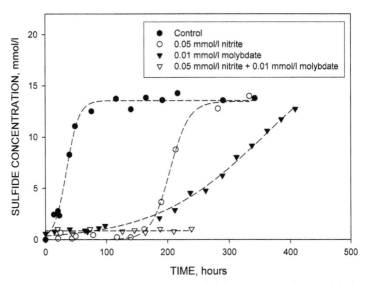

Figure 3.34 Effect of using nitrite and molybdate compounds on sulphide formation by a pure culture of *Desulfovibrio* sp. sulphate reducing bacteria [165].

Material factors

Attack of concrete by sulphuric acid differs somewhat from other forms of acid attack in that a low water/cement ratio is usually observed to increase rates of mass loss. Whilst the superficial response to this is to conclude that concrete exposed to attack by sulphur bacteria should have high water/cement ratios, the reality is, in fact, more complex.

The first point to note on this issue is that there may be differences in the behaviour of concrete exposed to sulphuric acid compared to concrete exposed to sulphur bacteria. This is illustrated in Figure 3.10, where a lower water/cement ratio appears to provide greater protection in experiments in which the concrete specimen is held in an environment containing sulphur bacteria. It should be noted that the concrete with the lower water/cement ratio in these experiments also had a slightly lower cement content (see discussion below). However, it is also conceivable that the lower ratio is limiting the extent to which water is absorbed by the concrete surface, providing a less habitable environment for sulphur oxidizing bacteria, as discussed in Section 3.4.2.

The second point is that concrete mix design conventions typically achieve lower water/cement ratios through an increase in cement content. Thus, in most of the studies in which the effect of water/cement ratio has been explored, the concrete mixes are, at least in part, more vulnerable to attack from sulphuric acid as a result of a higher quantity of calcium available to form ettringite, but more importantly gypsum. Thus, it is likely that greater protection of concrete can be realised through the reduction of water/cement ratio without increasing cement content. Water in concrete plays two fundamental roles—it acts as one of the reactants in the hydration of cement, and also as the fluid component which defines the ease with which the fresh material flows and compacts (its 'consistency' or 'workability'). If water/cement ratios are to be reduced without increasing cement content, this means a reduction in water content which may reduce the consistency of the mix unacceptably. Thus, it is likely that water-reducing admixtures (plasticizers and super-plasticizers) are required to play a role in achieving this goal. One study which has explored this approach found that a reduced water/cement ratio with constant cement content did, indeed, have greater resistance to attack from sulphuric acid [74].

There is, however, a strong incentive for taking the above approach a step further by reducing rather than maintaining the cement content, since it is the calcium in the cement matrix which is responsible for the formation of gypsum. This also explains why the inclusion of siliceous fly ash and silica fume as part of the cement are effective in reducing sulphuric acid attack, and the inclusion of slag less so: siliceous fly ash and silica fume have low calcium contents and will consequently reduce the total calcium content of the cement.

The use of cement/asbestos fibre mixtures in place of concrete in sewer pipes has been demonstrated to impart greater durability, although the precise reason for this was not explored by the researchers [68]. Asbestos is a hazardous substance and its use in new pipes is now uncommon. However, the possibility of using other fibres is one that may be worth exploring.

Whilst the protection afforded to concrete by protective coatings is essentially similar regardless of the nature of the acidic substances involved, some surface protection technologies are specifically used in pipes, and so should be discussed here. In many parts of the world, the use of concrete sewer pipes with internal polymer linings has become common practice. Pipe linings are typically made from polymeric materials. Historically, this was normally PVC, but high density polyethylene (HDPE) is now commonly used. Such liners can be incorporated in precast concrete pipes or installed in cast *in situ* pipes.

The installation of liners inside pipes after construction can be achieved through spray-on and brush-applied formulations. One such liner is discussed in Section 3.6.2. Another means of installing liners *in situ* is using rehabilitation systems in which a felt liner is saturated with epoxy resin and fed into the pipe. A rubber bladder is then inserted into the liner and inflated to stick the liner against the sides of the pipe. The bladder is deflated and the liner left to cure.

An undesirable side-effect of the use of inert liners is that there is no longer a 'sink' for hydrogen sulphide. Where exposed concrete is undergoing attack from sulphur bacteria, sulphur is effectively being captured in the form of poorly-soluble gypsum. Thus, hydrogen sulphide levels will be higher in liner sewer systems, leading to issues relating to odour [164]. Whilst this is not reason enough to reject the use of liners, it does mean that there may be additional costs associated with hydrogen sulphide control.

3.6.2 General approaches to protection

Constituents and mix proportions

From the experimental results presented throughout this chapter, other than for attack from sulphur bacteria, a reduced water-cement ratio unambiguously enhances resistance to attack from acids deriving from bacterial activity. Not only does a lower water/cement ratio reduce the rate at which the acid-deteriorated front progresses into the material, but it also limits the extent to which bacterial communities can establish themselves on concrete surfaces. It is also probable that a higher residual strength in the acid affected layer imparted by a lower water/cement ratio will enhance resistance to abrasion where this type of deterioration is occurring simultaneously to bacterial attack.

A lower cement content will also, generally, reduce susceptibility to attack. It should be noted that cement possesses a finite capacity for neutralizing acids. Thus, where a relatively small and finite quantity of acid is present, a higher cement content may, in fact, provide greater resistance. However, by its nature, this is an unlikely scenario in the case of bacterial deterioration. It has also been proposed that the higher volume of aggregate that arises from a lower cement content also enhances resistance to acid attack in other ways. Firstly, because aggregate particles tend to act as barriers to crack growth, their presence in higher number leads to a reduced crack density in the acid-deteriorated layer [170]. Additionally, aggregate acts as a restraint to shrinkage. It has been stated previously that cement which has undergone decalcification is prone to shrinkage and cracking resulting from this. Thus, the presence of aggregate acts to limit the extent to which cracking occurs in this zone [103].

Aggregates containing carbonate minerals impart a significantly greater neutralizing capacity in comparison to largely inert siliceous aggregate.

It has also been seen already that the inclusion of GGBS, fly ash and silica fume will typically enhance resistance to acids whose main mode of attack is acidolysis (which is the case for most of the main organic acids produced by bacteria). Generally, GGBS provides the highest level of resistance and silica fume the least. Other cementitious materials used in combination with Portland cement appear to perform less well. High alumina cement is also more resistant to attack from both nitric acid and organic acids formed by bacteria. The hydration products of these cements are generally less susceptible to acidolysis (specifically, there is an absence of portlandite), and they also possess a higher acid-neutralization capacity.

Admixed biocides appear to be of limited value with regards to bacterial colonisation. One study has examined the potential of including fibres impregnated with the biocide Microban in concrete [171]. The study also evaluated the performance of concrete containing zeolites containing silver and copper whose antimicrobial characteristics are already established. The concrete specimens were initially exposed to a hydrogen-sulphide environment prior to being placed in a vessel containing sulphur oxidising bacteria and nutrients. In both cases, there was very little improvement in resistance against attack. Another study, using a biocide based on silver, copper and zinc (presumably not in the form of zeolites) also found little improvement in the performance of concrete specimens stored in a manhole in a sewer for 17 months [69].

Coatings

Coatings are employed against bacterial degradation processes to prevent acid coming into contact with the concrete surface, and subsequently penetrating further into the pores of the material. For this reason, the most appropriate surface coatings for these applications are organic pore-blocking coatings [172]. These are formulations based on organic polymers which are applied in a liquid form to the surface. The liquid subsequently sets leaving a continuous solid barrier at the concrete surface. Such formulations include epoxy resins (possibly modified with coal tar) and polyurethane, polyester, vinyl and acrylic polymer materials.

Evaluation of coatings for concrete for protection against attack from bacterial deterioration and acid attack in the literature is largely focused on resistance to sulphur bacteria and sulphuric acid. However, since the purpose of all such coatings is to prevent the penetration of acid below the surface, performance can be expected to be similar for other acids.

Figure 3.35 is a bar chart showing the mass loss from concrete specimens (taken from commercial sewer pipes) exposed to sulphur oxidising bacteria after initially being exposed to an H_2S-bearing environment [170]. The surfaces of the specimens had previously been treated with a range of surface treatments with potential to protect against sulphuric acid. These were a cementitious coating, an epoxy coating and a sprayable polyurethane lining intended for the treatment of internal sewer pipe surfaces created by first spraying the surface with an epoxy primer, followed by hot-spraying with polyurethane. Both the lining and epoxy coating performed extremely well in isolating the concrete from the acid produced by the bacteria. The cementitious coating, presumably because it was made from cement and was, thus, essentially as vulnerable to attack as the concrete itself, performed poorly.

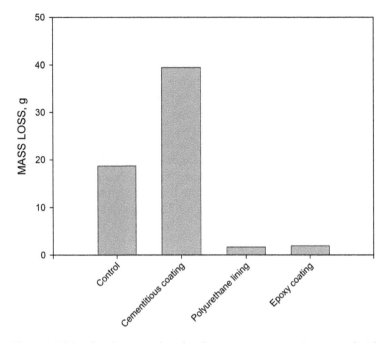

Figure 3.35 Mass loss from samples taken from concrete sewer pipes, treated with various surface coatings and exposed to sulphur oxidizing bacteria [170].

Measurements of the pH of concrete from the surface of concrete from sewer pipes found much less reduction in pH where the surface had been treated with epoxy resin, compared to untreated surfaces [173], thus indicating protection was achieved. Visual evaluation of concrete surfaces treated with epoxy and acrylic coatings and then exposed to a sulphuric acid solution with a pH of –0.9 concluded that the acrylic surface treatment was superior [174].

It should be stressed that the performance of surface treatments is very much dependent on the quality of workmanship during application. It is crucial that the concrete surface is in a condition suitable for receiving a coating, and this may mean that surface preparation is necessary. This may involve removal of surface contaminants and features deriving from the concrete itself (specifically laitance and efflorescence) which will compromise adhesion of the coating [172]. It may also involve filling blow holes at the surface and roughening the surface to enhance the bond. Roughening can be achieved through the use of wire brushes, impact tools, power grinding, sand/grit blasting or acid etching [175]. A primer coat may also require applying prior to the main coat or coats.

For most formulations, application is usually best carried out on a dry surface, but it is also important to ensure that there is as little moisture beneath the surface as possible, since this tends to compromise the bond between the concrete and the coating. Coatings can be applied using brushes, rollers and sprays.

Where abrasion of the concrete is also a possible deterioration mechanism, inorganic inclusions may be mixed with the organic coating to provide additional protection. This may include particles of sand or fibres. The cracking of concrete, resulting from loads in service or expansion and shrinkage resulting from wetting and drying, has the potential to severely compromise the effectiveness of coatings, since the coating may also crack, leaving an unprotected route into the interior of the concrete. Where this is likely to be an issue, coating formulation containing an elastomeric polymer (such as an elastomeric polyurethane) or fibres may provide some additional resilience [175]. One study examining the resistance of epoxy coatings reinforced with glass fibre mat found that, based on the performance of concrete specimens exposed to a 3% sulphuric acid solution, the presence of the coating was estimated to potentially extend service life by up to 70 times [176]. Where discontinuities ('holidays') in the coverage were present, however, performance was severely compromised.

It must be noted that surface coatings have a finite lifetime. Once a coating no longer performs its intended task it must be replaced. The longevity of coatings depends on the material used and the aggressive nature of the environment in which they are used, not just in terms of the aggressive substances that the coating is being used to provide protection against, but other factors such as UV exposure and the presence of oxidizing materials.

Guidance on concrete coatings has suggested that surface coatings can, if applied correctly, perform appropriately for periods of more than 15 years [172]. However, this may be optimistic for more aggressive environments, including many of those in which bacterial activity is present. A study has evaluated the durability of elastomeric coatings applied to surfaces of concrete—in which corrosion of steel had already initiated—to prevent the ingress of moisture and, thus, further corrosion [177]. The coatings were found to be effective for between 2 to 5 years. Whilst it should be stressed that this study did not examine bacterial attack or acid attack, these timescales still have relevance. One of the key mechanisms affecting the durability of coatings is the bond to the concrete substrate. The strength of this bond has been observed to decline with time in epoxy coatings, with the main mode of failure being localized debonding of the coating from the concrete (blistering), which can occur within two years of application [178].

Where longer-lasting protection is essential, the use of more substantial polymer liners is a more appropriate solution.

Polymer modification

Polymer modification of concrete involves the introduction of a polymer resin at the mixing stage which forms a matrix intermingled with that of a conventional inorganic cement. A number of different polymer types have been evaluated in terms of the level of acid resistance imparted to concrete, with varied results. For this reason, the findings of studies in this area are summarized in Table 3.10. The criterion used to judge success is, in most cases, mass loss. Beneficial effects have been observed for most polymer types, with the exception of acrylic polymers. Undesirable effects are sometimes also observed with polymer modification, including loss of

Table 3.10 Effectiveness of polymer modification with different polymer matrices.

	Study					
	[179]	[180]	[181]	[67]	[182]	[74]
Styrene acrylic ester	SB: improved	Lactic/ acetic acid: improved	SB: improved			
Acrylic polymer	SB: worsened		SB: worsened	SB: worsened NB: improved		
Styrene-butadiene	SB: improved		SB: improved			
Vinyl co-polymer	SB: improved		SB: improved			
Polysiloxane				SB: improved NB: improved		
Polycarboxylate				SB: improved NB: improved		
Polymethylmethacrylate					H_2SO_4: improved	
Polystyrene					H_2SO_4: improved	
Polyacrylonitrile					H_2SO_4: improved	
Polyvinyl acetate						H_2SO_4: improved

SB = sulphur bacteria; NB = nitrifying bacteria

strength, primarily due to retardation of cement hydration by the polymer [183].

Other studies have been conducted in which polymer modification has been conducted in combination with the use of other mix constituents intended to enhance acid resistance. This makes it difficult to evaluate which material is most important in imparting greater resistance. In one instance, fly ash and polyester resins were used in combination [184] yielding greater resistance to sulphuric, acetic, formic and lactic acid. In another study, concrete mixes made with a 'hybrid modified' formulation was evaluated [185]. 'Hybrid' in this instance means a cement matrix made from Portland cement, fly ash, polyvinyl acetate latex, and sodium silicate (Na_2SiO_3). In addition, a smaller quantity of sodium fluorosilicate (Na_2SiF_6) was added as a hardener for the sodium silicate. Deterioration was measured in terms of loss of compressive strength after exposure to a periodically refreshed solution of 1% sulphuric acid. The hybrid concrete performed better than the control mix, which contained the same quantity of fly ash, but without the latex and silicate compounds.

3.7 References

[1] Bennett PC, Rogers JR, Choi WJ and Hiebert FK (2001) Silicates, silicate weathering, and microbial ecology. Geomicrobiology Journal **18(1):** 3–19.

[2] Liu W, Xu X, Wu X, Yang Q, Luo Y and Christie P (2006) Decomposition of silicate minerals by *Bacillus mucilaginosus* in liquid culture. Environmental Geochemistry and Health **28(1-2):** 133–140.

[3] Hiebert FK and Bennett PC (1992) Microbial control of silicate weathering in organic-rich ground water. Science **258(5080):** 278–281.

[4] Friedrich S, Platonova NP, Karavaiko GI, Stichel E and Glombitza F (1991) Chemical and microbiological solubilization of silicates. Acta Biotechnologica **11(3):** 187–196.

[5] Duff RB, Webley DM and Scott RO (1963) Solubilization of minerals and related materials by 2-ketogluconic acid-producing bacteria. Soil Science **95(2):** 105–114.

[6] Lian B, Chen Y, Zhu L and Yang R (2008) Effect of microbial weathering on carbonate rocks. Earth Science Frontiers **15(6):** 90–99.

[7] Subrahmanyam G, Vaghela R, Bhatt NP and Archana G (2012) Carbonate-dissolving bacteria from 'miliolite', a bioclastic limestone, from Gopnath, Gujarat, Western India. Microbes and Environments **27(3):** 334–337.

[8] Subrahmanyam G (2014) Bacterial diversity and activity of semiarid soils of Mahi river basin Western India. Ph.D. Thesis, Maharaja Sayajirao University of Baroda, India, 346 p.

[9] Li W, Yu L-J, Wu Y, Jia L-P and Yuan DX (2007) Enhancement of Ca^{2+} release from limestone by microbial extracellular carbonic anhydrase. Bioresource Technology **98(4):** 950–953.

[10] Howell JA (1983) Mathematical models in microbiology: mathematical toolkit. *In*: Bazin M (ed.). Mathematics in Microbiology. Academic Press, New York, NY, USA.

[11] Viessman W, Hammer MJ, Perez EM and Chadik PA (2009) Water Supply and Pollution Control. 8th Ed., Pearson, Upper Saddle River, NJ, USA.

[12] Lewandowski Z (2000) Structure and function of biofilms. pp. 1–17. *In*: Evans LV (ed.). Biofilms: Recent Advances in their Study and Control. Harwood Acdemic Publishers, Amsgterdam, The Netherlands.

[13] DeBeer D, Stoodley P, Roe F and Lewandowski Z (1994) Effects of biofilm structures on oxygen distribution and mass transport. Biotechnology and Bioengineering **43(11):** 1131–1138.

[14] Elasri MO and Miller RV (1999) Study of the response of a biofilm bacterial community to UV radiation. Applied and Environmental Microbiology **65(5):** 2025–2031.

[15] Freeman C and Lock MA (1995) The biofilm polysaccharide matrix: A buffer against changing organic substrate supply? Limnology and Oceanography **40(2):** 273–278.

[16] Geesey GG, Jang L, Jolley JG, Hankins MR, Iwaoka T and Griffiths PR (1988) Binding of metal ions by extracellular polymers of biofilm bacteria. Water and Wastewater Microbiology **20(11-12):** 161–165.

[17] Fang HHP, Xu L-C and Chan K-Y (2002) Effects of toxic metals and chemicals on biofilm and biocorrosion. Water Research **36(19):** 4709–4716.

[18] Benzerara K, Menguy N, López-García P, Yoon T-H, Kazmierczak J, Tyliszczak T, Guyot F and Brown GE (2006) Nanoscale detection of organic signatures in carbonate microbialites. Proceeding of the National Academy of Science **103(25):** 9440–9445.

[19] Decho AW (2010) Overview of biopolymer-induced mineralization: What goes on in biofilms? Ecological Engineering **36(2):** 137–144.

[20] Bardy SL, Ng SYM and Jarrell KF (2003) Prokaryotic motility structures. Microbiology **149(2):** 295–304.

[21] Souza KA, Deal PH, Mack HM and Turnbill CE (1974) Growth and reproduction of microorganisms under extremely alkaline conditions. Applied Microbiology **28(6):** 1066–1068.

[22] Currie RJ (1986) Carbonation depths in structural quality concrete. Building Research Establishment, Watford, UK.

[23] Buenfield NR and Newman JB (1984) The permeability of concrete in a marine environment. Magazine of Concrete Research **36(127):** 67–80.

[24] Hoła J, Sadowski Ł, Reiner J, Stach S (2015) Usefulness of 3D surface roughness parameters for nondestructive evaluation of pull-off adhesion of concrete layers. Construction and Building Materials **84:** 111–120.

[25] Garbacz A and Courard L (2006) Characterization of concrete surface roughness and its relation to adhesion in repair systems. Materials Characterization **56(4-5):** 281–289.

[26] Kinnari TJ, Esteban J, Zamora N, Fernandez R, López-Santos C, Yubero F, Mariscal D, Puertolas JA and Gomez-Barrena E (2010) Effect of surface roughness and sterilization on bacterial adherence to ultra-high molecular weight polyethylene. Clinical Microbiology and Infection **16(7):** 1036–1041.

[27] Truong VK, Lapovok R, Estrin YS, Rundell S, Wang JY, Fluke CJ, Crawford RJ and Ivanova EP (2010) The influence of nano-scale surface roughness on bacterial adhesion to ultrafine-grained titanium. Biomaterials **31(13):** 3674–3683.

[28] Riedewald F (2006) Bacterial adhesion to surfaces: the influence of surface roughness. PDA Journal of Pharmaceutical Science and Technology **60(3):** 164–171.

[29] Mitik-Dineva N, Wang J, Mocanasu RC, Stoddart PR, Crawford RJ and Ivanova EP (2008) Impact of nano-topography on bacterial attachment. Biotechnology Journal **3(4):** 536–544.

[30] Giannantonio DJ, Kurth JC, Kurtis K and Sobecky PA (2009) Molecular characterizations of microbial communities fouling painted and unpainted concrete structures. International Biodeterioration and Biodegradation **63(1):** 30–40.

[31] Barberousse H, Ruot B, Yéprémian C and Boulon G (2007) An assessment of façade coatings against colonisation by aerial algae and cyanobacteria. Building and Environment **42(7):** 2555–2561.

[32] Dubosc A, Escadeillas G and Blanc PJ (2001) Characterization of biological stains on external concrete walls and influence of concrete as underlying material. Cement and Concrete Research **31(11):** 1613–1617.

[33] McWhirter MJ, Bremer PJ, Lamont IL and McQuillan AJ (2003) Siderophore-mediated covalent bonding to metal (oxide) surfaces during biofilm initiation by Pseudomonas aeruginosa bacteria. Langmuir 19(9): 3575–3577.

[34] Vu B, Chen M, Crawford RJ and Ivanova EP (2009) Review bacterial extracellular polysaccharides involved in biofilm formation. Molecules 14(7): 2535–2554.

[35] Alcamo IE, Fundamentals of microbiology. Jones and Bartlett, London, 832 p.

[36] Wang D, Cullimore R, Hu Y and Chowdhury R (2011) Biodeterioration of asbestos cement (AC) pipe in drinking water distribution systems. International Biodeterioration and Biodegradation 65(6): 810–817.

[37] Badger MR and Price GD (2003) CO_2 concentrating mechanisms in cyanobacteria: molecular components, their diversity and evolution. Journal of Experimental Botany 54(383): 609–622.

[38] Singh SK, Sundaram S and Kishor K (2014) Photosynthetic Microorganisms. Mechanism for Carbon Concentration. Springer, London, UK.

[39] British Standards Institution, BS EN 197-1:2011 Cement. Composition, specifications and conformity for common cements. British Standards Institution, London, UK.

[40] Nurse RW (1952) The effect of phosphate on the constitution and hardening of Portland cement. Journal of Applied Chemistry 2(12): 708–716.

[41] Ifka T, Palou MT and Bazelová Z (2012) The influence of CaO and P_2O_5 of bone ash upon the reactivity and the burnability of cement raw mixtures. Ceramics-Silikáty 56(1): 76–84.

[42] Eštoková A and Palaščákováa L (2012) Content of chromium and phosphorus in cements in relation to the slovak cement eco-labelling. Chemical Engineering Transactions 26: 75–80.

[43] British Standards Institution (2012) BS EN 450-1:2012—Fly ash for concrete Part 1: Definition, specifications and conformity criteria. British Standards Institution, London, UK.

[44] Hubbard FH, Dhir RK and Ellis MS (1985) Pulverised-fuel ash for concrete: Compositional characterisation of United Kingdom PFA. Cement and Concrete Research 15(1): 185–198.

[45] Shehata MH and Thomas MDA (2006) Alkali release characteristics of blended cements. Cement and Concrete Research 36(6): 1166–1175.

[46] de Larrard F, Gorse J-F and Puch C (1992) Comparative study of various silica fumes as additives in high-performance cementitious materials. Materials and Structures 25(149): 265–272.

[47] Hobbs DW (1986) Deleterious expansion of concrete due to alkali-silica reaction: Influence of PFA and slag. Magazine of Concrete Research 38(1376): 191–205.

[48] Dyer TD (2014) Concrete Durability. CRC Press, Boca Raton, FL, USA.

[49] American Society for Testing and Materials, ASTM C150/C150M—Standard Specification for Portland Cement. American Society for Testing and Materials, West Conshohocken, Pennsylvania, USA.

[50] British Standards Institution (2006) BS EN 15167-1—Ground granulated blast furnace slag for use in concrete, mortar and grout—Part 1: Definitions, specifications and conformity criteria. British Standards Institution, London, UK.

[51] Ortega-Calvo JJ, Sanchez-Castillo PM, Hernandez-Marine M and Saiz-Jimenez C (1993) Isolation and characterization of epilithic chlorophytes and cyanobacteria from two Spanish Cathedrals (Salamanca and Toledo). Nova Hedwiga 57(1): 239–253.

[52] Tomaselli L, Lamenti G, Bosco M and Tiano P (2000) Biodiversity of photosynthetic micro-organisms dwelling on stone monuments. International Biodeterioration and Biodegradation 46(3): 251–258.

[53] Crispim CA, Gaylarde PM and Gaylarde CC (2003) Algal and cyanobacterial biofilms on calcareous historic buildings. Current Microbiology 46(2): 79–82.

[54] Crowther RF and Harkness N (1975) Anaerobic bacteria. pp. 65–91. *In*: Curds CR and Hawkes HA (eds.). Ecological Aspects of Used Water Treatment—The Organisms and their Ecology. Vol. 1. Academic Press, London, UK.

[55] Lin ES (2004) A Modelling Study of H_2S Absorption of Pure Water and in Rainwater, Ph.D. thesis, National University of Singapore, Singapore.

[56] Martell AE and Smith RM (2004) Critical Selected Stability Constants of Metal Complexes Database, Version 8.0 for Windows. National Institute of Standards and Technology, Gaithersburg, MD, USA. See http://www.nist.gov/srd/nist46.cfm (accessed 07/06/2016).

[57] Postgate JR (1984) The Sulphate-Reducing Bacteria. 2nd Ed., Cambridge University Press, Cambridge, UK.

[58] Roberts DJ, Nica D, Zu G and Davis JL (2002) Quantifying microbially induced deterioration of concrete: initial studies. International Biodeterioration & Biodegradation **49(4):** 227–234.

[59] Robertson LA and Kuenen JG (2005) The colorless sulfur bacteria. pp. 985–1011. *In*: Balows A, Truper HG, Dworking M, Harder W and Schleifer K-H (eds.). The Prokaryotes. 2nd Ed., Springer-Verlag, Heidelberg, Germany.

[60] Sand W (1987) Importance of hydrogen sulfide, thiosulfate, and methylmercaptan for growth of thiobacilli during simulation of concrete corrosion. Applied and Environmental Microbiology **53(7):** 1645–1648.

[61] Nica D, Davis JL, Kirby L, Zuo G and Roberts DJ (2000) Isolation and characterization of microorganisms involved in the biodeterioration of concrete in sewers. International Biodeterioration and Biodegradation **46(1):** 61–68.

[62] Milde K, Sand W, Wolff W and Bock E (1983) Thiobacilli of the corroded concrete walls of the Hamburg sewer system. Journal of General Microbiology **129(5):** 1327–1333.

[63] Hernandez M, Marchand EA, Roberts D and Peccia J (2002) *In situ* assessment of active Thiobacillus species in corroding concrete sewers using fluorescent RNA probes. International Biodeterioration & Biodegradation **49(4):** 271–276.

[64] Hughes DC (1985) Sulphate resistance of OPC, OPC/fly ash, and SRPC pastes: pore structure and permeability. Cement and Concrete Research **15(6):** 1003–1012.

[65] Hill J, Byars EA, Sharp JH, Lynsdale CJ, Cripps JC and Zhou Q (2003) An experimental study of combined acid and sulfate attack of concrete. Cement and Concrete Composites **25(8):** 997–1003.

[66] Jiang G, Wightman E, Donose BC, Yuan Z, Bond PL and Keller J (2014) The role of iron in sulfide induced corrosion of sewer concrete. Water Research **49:** 166–174.

[67] Stanaszek-Tomal E and Fiertak M (2015) Biological and chemical corrosion of cement materials modified with polymer. Bulletin of the Polish Academy of Sciences Technical Sciences **63(3):** 591–596.

[68] Hewayde E, Nehdi M, Allouche E and Nakhla G (2007) Effect of mixture design parameters and wetting-drying cycles on resistance of concrete to sulfuric acid attack. Journal of Materials in Civil Engineering **19(2):** 155–163.

[69] Alexander MG and Fourie C (2011) Performance of sewer pipe concrete mixtures with Portland and calcium aluminate cements subject to mineral and biogenic acid attack. Materials and Structures **44(1):** 313–330.

[70] Guadalupe M, Gutiérrez-Padilla D, Bielefeldt A, Ovtchinnikov S, Hernandez M and Silverstein J (2010) Biogenic sulfuric acid attack on different types of commercially produced concrete sewer pipes. Cement and Concrete Research **40(2):** 293–301.

[71] O'Connell M, McNally C and Richardson MG (2012) Performance of concrete incorporating GGBS in aggressive wastewater environments. Construction and Building Materials **27(1):** 368–374.

[72] Girardi F, Vaona W and Di Maggio R (2010) Resistance of different types of concretes to cyclic sulfuric acid and sodium sulfate attack. Cement and Concrete Composites **32(8):** 595–602.

[73] Roy DM, Arjunan P and Silsbee MR (2001) Effect of silica fume, metakaolin, and low-calcium fly ash on chemical resistance of concrete. Cement and Concrete Research **31(12):** 1809–1813.

[74] Fattuhi NI and Hughes BP (1988) Ordinary portland cement mixes with selected admixtures subjected to sulfuric acid attack. ACI Materials Journal **85(6):** 512–518.

[75] De Belie N, Monteny J, Beeldens A, Vincke E, Van Gemert D and Verstraete W (2004) Experimental research and prediction of the effect of chemical and biogenic sulfuric acid on different types of commercially produced concrete sewer pipes. Cement and Concrete Research **34(12):** 2223–2236.

[76] Chang Z, Song X, Munn R and Marosszeky M (2005) Using limestone aggregates and different cements for enhancing resistance of concrete to sulphuric acid attack. Cement and Concrete Research **35(8):** 1486–1494.

[77] Tamimi AK (1997) High performance concrete mix for an optimum protection in acidic conditions. Materials and Structures **30(3):** 188–191.

[78] Torii K and Kawamura M (1994) Effects of fly ash and silica fume on the resistance of mortar to sulfuric acid and sulfate attack. Cement and Concrete Research **24(2):** 361–370.

[79] Siad H, Mesbah HA, Khelafi H, Kamali-Bernard S and Mouli M (2010) Effect of mineral admixture on resistance to sulphuric and hydrochloric acid attacks in self-compacting concrete. Canadian Journal of Civil Engineering **37(3):** 441–449.

[80] Durning TA and Hicks C (1991) Using microsilica to increase concrete's resistance to aggressive chemicals. Concrete International **13(3):** 42–48.

[81] Mehta PK (1985) Studies on chemical resistance of low water/cement ratio concretes. Cement and Concrete Research **15(6):** 969–978.

[82] Kim H-S, Lee S-H and Moon H-Y (2007) Strength properties and durability aspects of high strength concrete using Korean metakaolin. Construction and Building Materials **21(6):** 1229–1237.

[83] Sersale R, Frigione G and Bonavita L (1998) Acid depositions and concrete attack: main influences. Cement and Concrete Research **28(1):** 19–24.

[84] Türkel S, Felekoğlu B and Dulluç S (2007) Influence of various acids on the physico–mechanical properties of pozzolanic cement mortars. Sadhana **32(6):** 683–691.

[85] Adesanya DA and Raheem AA (2010) A study of the permeability and acid attack of corn cob ash blended cements. Construction and Building Materials **24(3):** 403–409.

[86] Scrivener KL, Cabiron J-L and Letourneux R (1999) High-performance concretes from calcium aluminate cements. Cement and Concrete Research **29(8):** 1215–1223.

[87] Geoffroy VA, Bachelet M and Croisier JL (2008) Evaluation of aluminium sensitivity of bacteria on a biodegrading *Acidithiobacillus Thiooxidans*: Definition of a specific growth medium. *In*: Calcium Aluminate Cements: Proceedings of the Centenary Conference, Avignon, 30 June–2 July 2008. IHS BRE Press, Garston, UK, pp. 309–331.

[88] Jariyathitipong P, Hosotani K, Fujii T and Ayano T (2014) Sulfuric acid resistance of concrete with blast furnace slag fine aggregate. Journal of Civil Engineering and Architecture **8(11):** 1403–1413.

[89] Thornton HT (1978) Acid attack of concrete caused by sulfur bacteria action. ACI Journal **75(11):** 577–584.

[90] Bock E, Koops H-P and Harms H (1986) Cell biology of nitrifying bacteria. *In*: Proser JI (ed.). Nitrification. IRL Press, Oxford, UK.

[91] Lipponen MTT, Martikainen PJ, Vasara RE, Servomaa K, Zacheus O and Kontro MH (2004) Occurrence of nitrifiers and diversity of ammonia-oxidizing bacteria in developing drinking water biofilms. Water Research **38(20):** 4424–4434.

[92] Gieseke A, Bjerrum L, Wagner M and Amann R (2003) Structure and activity of multiple nitrifying bacterial populations co-existing in a biofilm. Environmental Microbiology **5(5):** 355–369.

[93] Lydmark P, Lind M, Sörensson F and Hermansson M (2006) Vertical distribution of nitrifying populations in bacterial biofilms from a full-scale nitrifying trickling filter. Environmental Microbiology **8(11):** 2036–2049.

[94] Coskuner G and Curtis TP (2002) *In situ* characterization of nitrifiers in an activated sludge plant: Detection of *Nitrobacter* spp. Journal of Applied Microbiology **93(3):** 431–473.

[95] Eichler S, Christen R, Höltje C, Westphal P, Bötel J, Brettar I, Mehling A and Höfle MG (2006) Composition and dynamics of bacterial communities of a drinking water supply system as assessed by RNA- and DNA-based 16S rRNA gene fingerprinting. Applied and Environmental Microbiology **72(3):** 1858–1872.

[96] Regan JM, Harrington GW, Baribeau H, De Leon R and Noguera DR (2003) Diversity of nitrifying bacteria in full-scale chloraminated distribution systems. Water Research **37(1):** 197–205.

[97] Siripong S and Rittmann BE (2007) Diversity study of nitrifying bacteria in full-scale municipal wastewater treatment plants. Water Research **41(5):** 1110–1120.

[98] Meincke M, Krieg E and Bock E (1989) *Nitrosovibrio* spp., the dominant ammonia-oxidizing bacteria in building sandstone. Applied and Environmental Microbiology **55(8):** 2108–2110.

[99] Sand W, Ahlers B and Bock E (1991) The impact of microorganisms, especially nitric acid producing bacteria, on the deterioration of natural stones. pp. 481–484. *In*: Baer NS, Sabbioni C and Sors AI (eds.). Science, Technology, and European Cultural Heritage: Proceedings of the European Symposium, Bologna, Italy, 13–16 June 1989. Butterworth-Heinemann, Guildford, UK.

[100] Schön GH and Engel H (1962) Der einfluß des lichtes auf *Nitrosomonas europaea* win. Archiv für Mikrobiologie **42(4):** 415–428.

[101] Pavlík V (1994) Corrosion of hardened cement paste by acetic and nitric acids; Part II: formation and chemical composition of the corrosion products layer. Cement and Concrete Research **24(8):** 1495–1508.

[102] Pavlík V (1996) Corrosion of hardened cement paste by acetic and nitric acids Part III: Influence of water/cement ratio. Cement and Concrete Research **26(3):** 475–490.

[103] Pavlík V and Unčik S (1997) The rate of corrosion of hardened cement pastes and mortars with additive of silica fume in acids. Cement and Concrete Research **27(11):** 1731–1745.

[104] Khedr SA and Abou-Zeid MN (1994) Characteristics of silica-fume concrete. Journal of Materials in Civil Engineering **6(3):** 357–375.

[105] George CM (1997) Durability of calcium aluminate cement concrete: Understanding the evidence. pp. 253–263. *In*: Scrivener KL and Young JF (eds.). Mechanisms of Chemical Degradation of Cement-based Systems. CRC Press, Boca Raton, FL, USA.

[106] Sand W, Ahlers B and Bock E (2013) The impact of microorganisms—especially nitric acid producing bacteria—on the deterioration of natural stones. pp. 481–484. *In*: Baer NS, Sabbioni C and Sors AI (eds.). Science, Technology and European Cultural Heritage: Proceedings of the European Symposium, Bologna, Italy, 13–16 June 1989. Butterworth-Heinemann, Oxford, UK.

[107] Lyalikova, NN and Petushkova JP (1988) Mikrobiologicheskoe povrejdenie nastennoi jivopisi. Priroda **6:** 31–37.

[108] Sand W and Bock E (1991) Biodeterioration of mineral materials by microorganisms—biogenic sulfuric and nitric acid corrosion of concrete and natural stone. Geomicrobiology Journal **9(2-3):** 129–138.

[109] Jiang QQ and Bakken LR (1999) Comparison of *Nitrosospira* strains isolated from terrestrial environments. FEMS Microbiology Ecology **30(2):** 171–186.

[110] Sharma B and Ahler RC (1977) Nitrification and nitrogen removal. Water Research **11(10):** 897–925.

[111] Wasserbauer R, Zadák Z and Novotný J (1988) Nitrifying bacteria on the asbestos-cement roofs of stable buildings. International Biodeterioration **24(3):** 153–165.

[112] Muck R (2008) Silage Inoculants: What the Research Indicates about When and How to Use Them. US Department of Agriculture, Agricultural Research Service, Washington DC, USA.

[113] Bertron A and Duchesne J (2013) Attack of cementitious materials by organic acids in agricultural and agrofood effluents. pp. 131–173. *In*: Alexander M, De Belie N and Bertron A (eds.). Performance of Cement-based Materials in Aggressive Aqueous Environments, RILEM State-of-the-Art Report TC 211-PAE. Springer, Dordrecht, Netherlands,

[114] Yang K, Yu Y and Hwang S (2003) Selective optimization in thermophilic acidogenesis of cheese-whey wastewater to acetic and butyric acids: partial acidification and methanation. Water Research 37(10): 2467–2477.

[115] Johansen AG, Vegarud GE and Skeie S (2002) Seasonal and regional variation in the composition of whey from Norwegian Cheddar-type and Dutch-type cheeses. International Dairy Journal 12(7): 621–629.

[116] Larreur-Cayol S, Bertron A and Escadeillas G (2011) Degradation of cement-based materials by various organic acids in agro-industrial waste-waters. Cement and Concrete Research 41(8): 882–892.

[117] Bertron A, Duchesne J and Escadeillas G (2005) Attack of cement pastes exposed to organic acids in manure. Cement and Concrete Composites 27(9-10): 898–909.

[118] Dyer TD (2017) Influence of cement type on resistance to attack from two carboxylic acids. Cement and Concrete Composites (in press).

[119] Dyer TD (2017) Influence of cement type on resistance to organic acids. Magazine of Concrete Research 69(4): 175–200.

[120] Bertron A, Duchesne J and Escadeillas G (2005) Attack of cement pastes exposed to organic acids in manure. Cement and Concrete Composites 27(9-10): 898–909.

[121] De Belie N, Debruyckere M, Van Nieuwenburg D and De Blaere B (1997) Concrete attack by feed acids: accelerated tests to compare different concrete compositions and technologies. ACI Materials Journal 94(6): 546–554.

[122] De Belie N, De Coster V and Van Nieuwenburg D (1997) Use of fly ash or silica fume to increase the resistance of concrete to feed acids. Magazine of Concrete Research 49(181): 337–344.

[123] De Belie N, Verselder HJ, De Blaere B, Van Nieuwenburg D and Verschoore R (1996) Influence of the cement type on the resistance of concrete to feed acids. Cement and Concrete Research 26(11): 1717–1725.

[124] Bertron A, Duchesne J and Escadeillas G (2005) Accelerated tests of hardened cement pastes alteration by organic acids: Analysis of the pH. Cement and Concrete Research 35(1): 155–166.

[125] Bertron A, Duchesne J and Escadeillas G (2007) Degradation of cement pastes by organic acids. Materials and Structures 40(3): 341–354.

[126] Kawai K, Teranishi S, Morinaga T and Tazawa EI (1993) Concrete deterioration caused by anaerobic bacteria. Translation form Journal of Materials, Concrete Structures and Pavements 21(478) in Concrete Library of JSCE, 1994, pp. 127–139.

[127] Maruthamuthu S, Saraswathi V, Mani A, Kalyanasundaram RM and Rengaswamy NS (1997) Influence of freshwater heterotrophic bacteria on reinforced concrete. Biofouling 11(4): 313–323.

[128] Parande AK, Muralidharan S, Saraswathy V and Palaniswamy N (2005) Influence of microbiologically induced corrosion of steel embedded in ordinary Portland cement and Portland pozzolona cement. Anti-Corrosion Methods and Materials 52(3): 148–153.

[129] Gaylarde CC and Gaylarde PM (2005) A comparative study of the major microbial biomass of biofilms on exteriors of buildings in Europe and Latin America. International Biodeterioration and Biodegradation 55(2): 131–139.

[130] Barberousse H, Lombardo RJ, Tell G and Couté A (2006) Factors involved in the colonisation of building façades by algae and cyanobacteria in France. Biofouling 22(2): 69–77.

[131] Ruttner F (1963) Fundamentals of Limnology (translated by Frey DG and Fry FEJ). University of Toronto Press, Toronto, Canada.

[132] Terzaghi RD (1948) Concrete deterioration in a shipway. American Concrete Institute Proceedings 44(6): 977–1005.

[133] Ballim Y and Alexander MG (1990) Carbonic acid water attack of Portland cement based matrices. *In*: Dhir R and Green J (eds.). Protection of Concrete: Proceedings of the International Conference, University of Dundee, September 1990. Spon, London, UK.

[134] Colasanti R, Coutts D, Pugh SYR and Rosevear A (1991) The microbiology programme for UK Nirex. Experientia **47(6)**: 560–572.

[135] Bayer EA, Shoham Y and Lamed R (2006) Cellulose-decomposing bacteria and their enzyme systems. pp. 578–617. *In*: Dworkin M, Falkow S, Rosenberg E, Schleifer K-H and Stackebrandt E (eds.). The Prokaryotes: Volume 2: Ecophysiology and Biochemistry. Springer, New York, NY, USA.

[136] Walczak I, Libert M-F, Camaro S and Blanchard J-M (2001) Quantitative and qualitative analysis of hydrosoluble organic matter in bitumen leachates. Agronomie **21(3)**: 247–257.

[137] Libert M-F and Walczak I (2000) Effect of radio-oxidative ageing and pH on the release of soluble organic matter from bitumen. *In*: Proceedings of ATALANTE 2000 Scientific Research on the Back-End of the Fuel Cycle for the 21st Century. Avignon, France, pp. 1–4.

[138] Nakayama S, Iida Y, Nagano T and Akimoto T (2003) Leaching behavior of a simulated bituminized radioactive waste form under deep geological conditions. Journal of Nuclear Science and Technology **40(4)**: 227–237.

[139] Bertron A, Jacquemet N, Erable B, Sablayrolles C, Escadeillas G and Albrecht A (2014) Reactivity of nitrate and organic acids at the concrete–bitumen interface of a nuclear waste repository cell. Nuclear Engineering and Design **268**: 51–57.

[140] Arter HE, Hanselmann KW and Bachofen R (1991) Modelling of microbial degradation process: the behaviour of microorganisms in a waste repository. Experientia **47(6)**: 578–584.

[141] Pedersen K (1999) Subterranean microorganisms and radioactive waste disposal in Sweden. Engineering Geology **52(3-4)**: 163–176.

[142] MacDonald DD, Roberts B and Hyne JB (1978) The corrosion of carbon steel by wet elemental sulfur. Corrosion Science **18(5)**: 411–425.

[143] Boden PJ and Maldonado-Zagal SB (1982) Hydrolysis of elemental sulfur in water and its effects on the corrosion of mild steel. British Corrosion Journal **17(3)**: 116–120.

[144] Nemati M, Jenneman GE and Voordouw G (2001) Impact of nitrate-mediated microbial control of souring in oil reservoirs on the extent of corrosion. Biotechnology Progress **17(5)**: 852–859.

[145] Kodama Y and Watanabe K (2004) *Sulfuricurvum kujiense* gen. nov., sp. nov., a facultatively anaerobic, chemolithoautotrophic, sulfur-oxidizing bacterium isolated from an underground crude-oil storage cavity. International Journal of Systematic and Evolutionary Microbiology **54(6)**: 2297–2300.

[146] Oil and Gas UK (2015) Guidelines on Qualification of Materials for the Abandonment of Wells, Issue 2, Oil and Gas UK, Aberdeen, UK.

[147] Lundegard PD and Kharaka YK (1994) Distribution and occurrence of organic acids in subsurface waters. *In*: Pittman ED and Lewan MD (eds.). Organic Acids in Geological Processes. Springer-Verlag, London, UK.

[148] US Environmental Protection Agency (1985) Design Manual—Odor and Corrosion Control in Sanitary Sewerage Systems and Treatment Plants. US Environmental Protection Agency, Office of Research and Development, Cincinnati, OH, USA.

[149] US Environmental Protection Agency (1991) Technical Report 430/09-91-010—Hydrogen Sulfide Corrosion in Wastewater Collection and Treatment Systems. US Environmental Protection Agency, Washington, DC, USA.

[150] Pomeroy RD and Parkhurst RD (1977) The forecasting of sulfide build-up rates in sewers. Progress in Water Technology **9(3)**: 621–628.

[151] Stewart PS, Rayner J, Roe F and Rees WM (2001) Biofilm penetration and disinfection efficacy of alkaline hypochlorite and chlorosulfamates. Journal of Applied Microbiology **91(3)**: 525–532.

[152] Tanaka N, Hvitved-Jacobsen T, Ochi T and Sato N (2000) Aerobic–anaerobic microbial wastewater transformations and re-aeration in an air-injected pressure sewer. Water Environment Research 72(6): 665–674.

[153] Garcia-de-Lomas J, Corzo A, Portillo MC, Gonzalez JM, Andrades JA, Saiz-Jimenez C and Garcia-Robledo E (2007) Nitrate stimulation of indigenous nitrate-reducing, sulfide-oxidising bacterial community in wastewater anaerobic biofilms. Water Research 41(14): 3121–3131.

[154] Jiang G, Sharma KR and Yuan Z (2013) Effects of nitrate dosing on methanogenic activity in a sulfide-producing sewer biofilm reactor. Water Research 47(5): 1783–1792.

[155] Heukelelekian H (1943) Effect of the addition of sodium nitrate to sewage on hydrogen sulfide production and B.O.D. reduction. Sewage Works 15(2): 255–261.

[156] Szabo JG, Rice EW and Bishop PL (2007) Persistence and decontamination of *Bacillus atrophaeus* subsp. *globigii* spores on corroded iron in a model drinking water system. Applied Environmental Microbiology 73(8): 2451–2457.

[157] Morrow JB, Almeida JL, Fitzgerald LA and Cole KD (2008) Association and decontamination of Bacillus spores in a simulated drinking water system. Water Research 42(20): 5011–5021.

[158] Firer D, Friedler E and Lahav O (2008) Control of sulfide in sewer systems by dosage of iron salts: Comparison between theoretical and experimental results, and practical implications. Science of the Total Environment 392(1): 145–156.

[159] Zhang L, De Schryver P, De Gusseme B, De Muynck W, Boon N and Verstraete W (2008) Chemical and biological technologies for hydrogen sulfide emission control in sewer systems: A review. Water Research 42(1–2): 1–12.

[160] Gutierrez O, Park D, Sharma KR and Yuan Z (2009) Effects of long-term pH elevation on the sulfate-reducing and methanogenic activities of anaerobic sewer biofilms. Water Research 43(9): 2549–2557.

[161] Othman F, MortezaNia S, Ghafari S and Hashim S (2011) Suppressing dissolved hydrogen sulfide in a sewer network using chemical methods. Scientific Research and Essays 6(17): 3601–3608.

[162] Gutierrez O, Sudarjanto G, Ren G, Ganigué R, Jiang G and Yuan Z (2014) Assessment of pH shock as a method for controlling sulfide and methane formation in pressure main sewer systems. Water Research 48: 569–578.

[163] United States Environmental Protection Agency (1999) Collection Systems—O&M Fact Sheet—Sewer Cleaning and Inspection. United States Environmental Protection Agency, Washington, DC, USA.

[164] Nielsen AH, Vollertsen J, Jensen HS, Wium-Andersen T and Hvitved-Jacobsen T (2008) Influence of pipe material and surfaces on sulfide related odor and corrosion in sewers. Water Research 42(15): 4206–4214.

[165] Nemati M, Mazutinec TJ, Jenneman GE and Voordouw G (2001) Control of biogenic H₂S production with nitrite and molybdate. Journal of Industrial Microbiology and Biotechnology 26(6): 350–355.

[166] Telang AJ, Ebert S, Foght JM, Westlake DWS and Voordouw G (1998) Effects of two diamine biocides on the microbial community from an oil field. Canadian Journal of Microbiology 44(11): 1060–1065.

[167] Reinsel MA, Sears JT, Stewart PS and McInerney MJ (1996) Control of microbial souring by nitrate, nitrite or glutaraldehyde injection in a sandstone column. Journal of Industrial Microbiology 17(2): 128–136.

[168] Predicala B, Nemati M, Stade S and Laguë C (2008) Control of H₂S emission from swine manure using Na-nitrite and Na-molybdate. Journal of Hazardous Materials 154(1-3): 300–309.

[169] Jayaraman A, Mansfeld FB and Wood TK (1999) Inhibiting sulfate-reducing bacteria in biofilms by expressing the antimicrobial peptides indolicidin and bactenecin. Journal of Industrial Microbiology and Biotechnology 22(3): 167–175.

[170] Fattuhi NI and Hughes BP (1988) Ordinary Portland cement mixes with selected admixtures subjected to sulfuric acid attack. ACI Materials Journal **85(6):** 512–518.

[171] De Muynck W, De Belie N and Verstraete W (2009) Effectiveness of admixtures, surface treatments and antimicrobial compounds against biogenic sulfuric acid corrosion of concrete. Cement and Concrete Composites **31(3):** 163–170.

[172] Concrete Society (1997) Concrete Society Technical Report No. 50—Guide to Surface Treatments for Protection and Enhancement of Concrete. Concrete Society, Slough, UK.

[173] Valix M, Zamri D, Mineyama H, Cheung WH, Shi J and Bustamante H (2012) Microbiologically induced corrosion of concrete and protective coatings in gravity sewers. Chinese Journal of Chemical Engineering **20(3):** 433–438.

[174] Aguiar JB, Camões A and Moreira PM (2008) Coatings for concrete protection against aggressive environments. Journal of Advanced Concrete Technology **6(1):** 243–250.

[175] Roy SK (2002) Coated Concrete—Why, What and How. *In:* Bassi R and Roy SK (eds.). Handbook of Coatings for Concrete. Whittles, Caithness, UK.

[176] Vipulanandan C and Liu J (2002) Glass-fiber mat-reinforced epoxy coating for concrete in sulfuric acid environment. Cement and Concrete Research **32(2):** 205–210.

[177] Seneviratne A, Sergi G and Page C (2000) Performance characteristics of surface coatings applied to concrete for control of reinforcement corrosion. Construction and Building Materials **14(1):** 55–59.

[178] Liu J and Vipulanandan C (2005) Tensile bond strength of epoxy coatings to concrete substrate. Cement and Concrete Research **35(7):** 1412–1419.

[179] Vincke E, Van Wanseele E, Monteny J, Beeldens A, De Belie N, Taerwe L, Van Gemert D and Verstraete W (2002) Influence of polymer addition on biogenic sulfuric acid attack of concrete. International Biodeterioration and Biodegradation **49(4):** 283–292.

[180] De Belie N and Monteny J (1998) Resistance of concrete containing styrol acrylic acid ester latex to acids occurring on floors for livestock housing. Cement and Concrete Research **28(11):** 1621–1628.

[181] Monteny J, De Belie N, Vincke E, Verstraete W and Taerwe L (2001) Chemical and microbiological tests to simulate sulfuric acid corrosion of polymer-modified concrete. Cement and Concrete Research **31(9):** 1359–1365.

[182] Bhattacharya VK, Kirtania KR, Maiti MM and Maiti S (1983) Durability tests on polymer-cement mortar. Cement and Concrete Research **13(2):** 287–290.

[183] Beeldens A, Monteny J, Vincke E, De Belie N, Van Gemert D, Taerwe L and Verstaete W (2001) Resistance to biogenic sulphuric acid corrosion of polymer-modified mortars. Cement and Concrete Composites **23(1):** 47–56.

[184] Gorninski JP, Dal Molin DC and Kazmierczak CS (2007) Strength degradation of polymer concrete in acidic environments. Cement and Concrete Composites **29(8):** 637–645.

[185] Li G, Xiong G, lü Y and Yin Y (2009) The physical and chemical effects of long-term sulphuric acid exposure on hybrid modified cement mortar. Cement and Concrete Composites **319(5):** 325–330.

Chapter 4

Fungal Biodeterioration

4.1 Introduction

Fungi are eukaryotes: their cells contain nuclei and organelles both of which are enclosed by membranes. This configuration of the cell is the same as that for plants and animals.

Species of fungi can live in terrestrial or aquatic environments, including seawater. Superficially fungi may seem very like plants—both include unicellular and multicellular organisms, lack motility, and many of the multicellular forms share some common morphological features. However, in taxonomic terms, fungi occupy a separate kingdom to these other organisms. There are a number of reasons for this. Firstly, unlike plants, fungi obtain energy and carbon from organic compounds in the surrounding environment. Secondly, the cell walls of fungi contain the protein chitin, as opposed to plants, whose cell walls are cellulose.

Fungi can take the form of both single and multicellular organisms. It is useful to discuss these forms separately, as the differences have some significance to some of the topics covered later in this chapter.

4.1.1 Yeasts

Yeasts differ from other forms of fungi both structurally and in their means of reproducing. Yeasts are unicellular, like bacteria, meaning that they are composed of single discrete cells. Unlike bacteria, yeasts are non-motile, and so only move when fluids in which they are present are mobile. Some yeast are capable of attaching to surfaces, which occurs through the formation of adhesive compounds, including sugars, at the cell surface [1].

Like bacteria, some yeasts reproduce asexually through fission. However, it is more common for asexual reproduction to occur via budding. Budding involves the formation of a small bud on the side of the yeast cell. Once formed, the nucleus of the cell also splits to form two identical nuclei (*mitosis*), with one half migrating into the bud. The bud then separates from the parent cell. Yeasts can also reproduce sexually, usually when the organism faces a shortage of nutrients or other stress.

4.1.2 Multicellular fungi

Multicellular fungi differ from yeast in that they grow in the form of a system of tubes which are capable of growth and branch formation (see Figure 4.1). These tubes are referred to as *hyphae*, with a grouping of hyphae referred to as the *mycelium*. The hyphae are effectively cells, although fungi vary in how these hyphae are structured—in some cases individual hyphae will be structured as a single cell, whilst others are divided up into multiple cells by walls known as *septa*. However, the septa are normally perforated, allowing fluid and organelles to move between the partitions.

Beyond this definition of multicellular fungi, there exists a great deal of structural variation, from relatively simple moulds and rusts, to complex structures including mushrooms and puffballs. Thus, further discussion of the physical form that such fungi take will be made as required in discussing those which affect concrete.

4.1.3 Lichen

Lichen are composite organisms containing more than one species, where what may be a symbiotic relationship exists. However, the mechanisms by

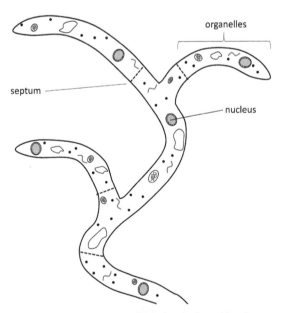

Figure 4.1 A schematic diagram of a fungal hypha.

which lichen may affect a concrete substrate on which they are growing are largely the result of the fungal component of this co-existence.

The organisms that constitute a lichen are fungi (the *mycobiont*) and photosynthesising organisms (the *photobiont*)—either algae (Chapter 5) or cyanobacteria (Chapter 3). Usually, there are two species involved, but in some cases three may be present [2].

Lichen adopt many different growth types, including *crustose* (forming relatively flat structures whose underside is fully attached to the substrate on which they grow), *foliose* (leaf-like) and *fruticose* (hair-like, strap-shaped or shrub-like) [3]. Of specific interest with regards to concrete durability is the *endolithic* growth form where the lichen exists entirely beneath a mineral surface [4].

The photobionts are embedded within the structure of the mycobiont, either evenly distributed through the body of the lichen (the *thallus*), or present as discrete layers.

The photobionts in lichens undergo photosynthesis and produce organic compounds for the mycobiont to use as a source of energy and carbon. It has been queried whether the relationship is truly symbiotic, on the grounds that there does not appear to be much benefit to the photobiont. Indeed, it has been suggested that the relationship is potentially parasitic in nature [2]. However, it is thought that residing within the fungus provides the photobiont with a degree of protection from extreme conditions that it would otherwise not have [6].

4.2 Fungal Metabolism

Fungi are heterotrophs, meaning that they obtain carbon and energy from organic compounds. The compounds that can be used by fungi must be relatively small molecules to allow them to be absorbed through the cell or hyphal wall. However, fungi are capable of breaking down large molecules into smaller ones through the use of enzymes. Enzymes are either released into the surrounding environment, or are present on the surface of the hyphae.

Many fungi feed on plant matter, which contains a significant proportion of cellulose and hemicellulose. These compounds are polymers whose monomeric units are sugars. Enzymes are used to cleave the bonds between the sugar molecules, and break down the polymers into smaller and smaller fragments until molecules such as glucose and fructose remain, which can be absorbed.

Fungi are either obligate aerobes (they require oxygen for the production of energy) or facultative anaerobes (not requiring oxygen, but with the means of using oxygen where it is available).

The anaerobic process will often yield ethanol, as in the ethanol fermentation carried out by yeasts in the production of wine and beer.

However, a range of other metabolites may be formed including citric, oxalic, gluconic, succinic, formic and malic acids [5].

The aerobic process of metabolising organic molecules leads to the formation of carbon dioxide and water:

$$C_6H_{12}O_6 + 6O_2 \rightarrow 6CO_2 + 6H_2O$$

In the above example, D-glucose is the organic compound involved.

4.3 Growth of Fungal Organisms and Communities

4.3.1 Growth

When nutrients are plentiful, fungi undergo exponential growth of the mycelium, which is referred to as the *trophophase* [6]. However, where a nutrient becomes scarce, *secondary metabolites* are formed whose purpose is, at least partly, to initiate processes which offer the possibility of resolving the nutrient shortage in a number of different ways. These metabolites can include organic acids—many similar to those formed by anaerobic processes—antibiotics and enzymes. At this point the trophophase ends and the *idiophase* begins. During the idiophase development of biomass slows or stops and the manner in which the hyphae grow changes to one in which the formation of new branches becomes dominant.

A wide range of organic acids are produced by fungi as secondary metabolites, including oxalic, citric, gluconic and malic acid. Those with greatest significance to concrete durability are citric and oxalic acid. There are a number of theories as to why these acids are formed.

In the case of oxalic acid, it is believed that the main reason for its production is to increase rates at which plant matter is broken down [7]. Oxalic acid production is normally stimulated by poor nutrient availability with respect to carbon [8]. The source of nitrogen may also be important, with at least one study finding oxalic acid production to be much higher when nitrogen was present as nitrate ions rather than ammonium ions [9]. Higher concentrations of calcium also tend to promote oxalic acid production [10].

Oxalic acid assists in the breakdown of plant matter for a number of reasons. Firstly, cellulose begins to break down at low pHs. Secondly, many of the enzymes employed by fungi to break down cellulose are most effective at low pHs. Thirdly, it would appear that oxalic acid plays a role in promoting the Fenton reaction, in which Fe(III) is reduced to Fe(II), leading to the production of hydroxyl radicals which are employed in attacking cellulose through oxidative degradation.

When oxalic acid comes into contact with calcium ions, calcium oxalate is precipitated. The formation of this compound by fungi can take two forms. Firstly oxalic acid and calcium come together within the walls of the hyphae to form crystals of calcium oxalate which begin to protrude through the walls into the external environment as they grow [11]. However, oxalic acid may also be released in exudates produced at the surfaces of the hyphae. In this case the acid molecules encounter calcium ions in the external environment and crystals are precipitated. Where calcium oxalate is produced from exudate, it has been proposed that this may act as a means of disrupting calcium pectate, found in cell walls of plants, allowing access to cell interiors [12].

Much of the understanding of citric acid formation by fungi arises from the industrial exploitation of *Aspergillus niger* as a means of manufacturing this compound. However, similar factors influence fungi in nature. Citric acid formation is promoted when certain nutrients become scarce. Of particular importance are nitrogen [13], phosphorus [14] and the metals iron, manganese and zinc [15]. Citric acid is also produced in larger quantities under pH conditions of less than 2 [16] and (within a bioreactor) at higher dissolved oxygen concentrations [17].

It has been proposed that the drop in pH associated with citric acid production creates a hostile environment for other organisms, meaning that acid producing fungi may gain an advantage over other bacteria and fungi. Many of the acids formed are also effective at forming strong complexes with metal ions, allowing them to solubilize nutrients that would otherwise be relatively insoluble, and break down mineral matrices to release trace quantities of nutrients held within [7].

As discussed earlier, lichen consist of photobionts (which undergo photosynthesis) and mycobionts which are heterotrophic fungi which obtain energy and carbon from compounds formed by the photobionts. Where the photobionts are algae, the compounds formed are the polyols, erythritol, ribitol and sorbitol, depending on the genus involved [6]. Where they are cyanobacteria, the compound is glucose. The process of photosynthesis is discussed in further detail in Chapter 5, and the manner in which the mycobiont in lichen processes these organic compounds is identical to that for other fungi. Lichen commonly produce oxalic acid as a secondary metabolite, but may also produce a range of other organic acids referred to collectively as lichenic acids. These are discussed later in this chapter.

Fungi require water, which is required to achieve the turgor pressures needed for hyphae to grow. In many cases fungi may be in contact with plentiful supplies of water, but where they are exposed to the atmosphere, approaches to retain water are adopted. Fungi living on surfaces frequently produce extracellular polymeric substances (EPS), to form biofilms similar to those of bacteria. These biofilms have good water retention capabilities [18].

In the case of fungi living in seawater, where the process of osmosis would otherwise cause the movement of water out of the hyphae and into the salt-rich water in the surrounding environment, this is done by synthesising high concentrations of water-soluble compounds known as polyols such that the solute concentration in the hyphae is higher than that in seawater [6]. Moulds which grow on salty or sugary foodstuffs, such as *Aspergillus eurotium* and various species of *Penicillium* produce glycerol, or absorb solutes from the surrounding environment to achieve the same effect.

Under desiccating conditions, lichens also produce polyols, which replace water molecules associated with macromolecules in their cells. This allows them to survive long periods of desiccation in a seemingly lifeless state, which they recover from on re-hydration [6].

Fungi, being heterotrophic do not undergo photosynthesis. Whilst they utilise light as a means of triggering processes such as spore production and dispersal, exposure to UV radiation is usually detrimental. For this reason, many fungi exist with much or all of their mycelium in the dark. For fungi growing on exposed mineral surfaces, one solution is to produce pigments such as melanins, carotenoids and mycosporines, which provide protection from sunlight [18]. As we will see later, another option on at least some mineral surfaces is to send hyphae beneath the surface.

4.3.2 Reproduction

During the idiophase, once the hyphae begin to encounter the edge of the substrate on which they are growing the process of reproduction may begin. This usually takes the form of the formation of spore forming organs, which can take many different forms.

Fungi are capable of both asexual and sexual reproduction, with both types of reproduction usually playing a role in their life-cycle. Asexual reproduction can occur through mycelial fragmentation where a mycelium splits into two separate mycelia which grow separately from each other—vegetative growth. More commonly, asexual reproduction occurs through spore formation. This involves the development of spore forming organs by the fungus. Hyphae in these organs undergo mitosis, and one of the nuclei is encapsulated within a spore which grows from the hyphae before detaching. The spore is released into the wider environment and may be dispersed by the wind, by flowing water, or on the surface or in the digestive system of animals. Some spores are motile. Once the spore is in a location conducive to growth it will begin to form hyphae and grow into a mycelium.

Sexual reproduction occurs when hyphae from two separate mycelia undergo meiosis to form haploid cells (cells containing a set of unpaired chromosomes) which become fused together. The process of fusion can take many forms, but the product is a zygotic diploid cell (containing paired chromosomes). This cell then undergoes meiosis to form spores, which are released. The important aspect of the spore formation process from the perspective of concrete durability is that spores can be travel widely, meaning that any concrete surface in contact with air or water is potentially capable of acquiring fungal occupants.

The fungal component of lichen can reproduce asexually or sexually in the same manner as other fungi. Where asexual reproduction is vegetative, both divided parts will contain photobionts. However, where reproduction occurs via spores, there is no transfer of photobiont to the new organisms. Thus, the mycobiont must normally reacquire photobionts, a process which, in most cases, appears to rely on chance confluence [19].

4.3.3 Fungal communities

Because enzymes are employed by fungi outside the organism, there is limited control over their formation of absorbable nutrients. This can often lead to the production of excess quantities. Sometimes, these nutrients are exploited by other fungi or bacteria, leading to the development of complex communities of microorganisms [20]. Conversely, some fungi release antibiotic compounds which inhibit growth and reproduction of other fungal species and other micro-organisms. Bacteria can also produce similar compounds, with the purpose of suppressing fungal activity.

4.4 Concrete as a Habitat for Fungal Life

In discussing the type of habitat a concrete surface provides to fungi, there exists some overlap between the requirements of fungus and those of bacteria. Where this is the case, the reader is referred back to Chapter 3. However, there are sufficient differences in these requirements to further discuss them. Moreover, research has been carried out into fungal colonization of concrete the results of which are useful in understanding how damage is done to concrete and also how this damage might be prevented. The three key aspects which require consideration remain the same—the issue of pH, the abundance or scarcity of the various nutrients required by fungi, and the physical habitat that a concrete surface provides fungi.

4.4.1 pH

Most fungi grow at optimal rates under slightly acidic conditions. This is illustrated in Figure 4.2 which shows the magnitude of fungal growth in soil samples from a strip of land over which a pH gradient exists. The optimum conditions for growth in this case are at a pH of around 5. The optimum pH for growth, however, is dependent on the substrate in which the fungus is growing.

Some fungus display optimal growth rates under alkaline conditions. Some coprinus species (*Coprinus radiatus, Coprinus micaceus* and *Coprinus ephemerus*) and carbonicolous ('coal-inhabiting') fungi have been found to display optimal growth rates under pH conditions close to 8 [22, 23].

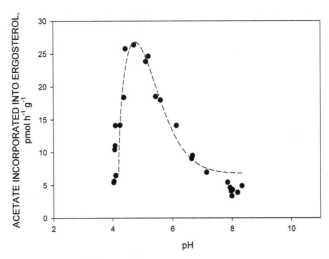

Figure 4.2 Growth of fungi (measured as acetate incorporation into ergosterol) in soil samples taken from a strip of soil in which a pH gradient exists [21].

Figure 4.3 shows quantities of fungal biomass developed by a range of ammonia fungi, plus some non-ammonia fungi over a wide pH range. Many of the early-phase ammonia fungi display optimum growth between pH 7 and 8. Some of the species still display relatively large growth rates at pH 9, with indications that growth may still be observed at still higher values. However, it should be stressed that, none of the fungal genera discussed above or in Figure 4.3 have been observed to colonise concrete surfaces.

Indeed, it is likely that in most cases, as is the case for bacteria (Chapter 3), carbonation of concrete is necessary before fungal colonization can occur. This has been demonstrated in a study in which concrete specimens were inoculated with three isolated fungal strains (*Alternaria alternata*, a species from the *Exophiala* genus, and *Coniosporium uncinatum*) and incubated at 26°C [25]. The specimens were made from the same white Portland cement paste, but one group had been 'weathered' through exposure to a carbonating atmosphere, whilst another was carbonated and then leached by de-ionised water. After 4 weeks, the non-weathered specimens had only small patches of fungal growth, whilst the weathered specimens showed considerably more advanced growth, with the carbonated and leached specimens showing the highest coverage by fungi (Figure 4.4).

Lichen are subdivided in terms of the pH of the substrate on which they are capable of growing. Silicicolous fungi require acidic substrates, whilst calcicolous grow on neutral to alkaline substrates [4].

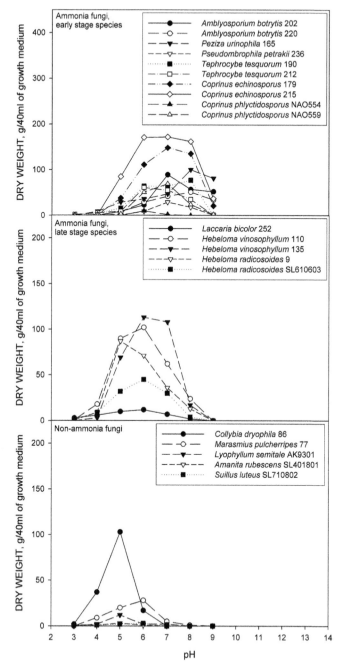

Figure 4.3 Biomass developed by a range of ammonia and non-ammonia fungi in growth media with adjusted pH, incubated in the dark at 23°C for 14 days [24].

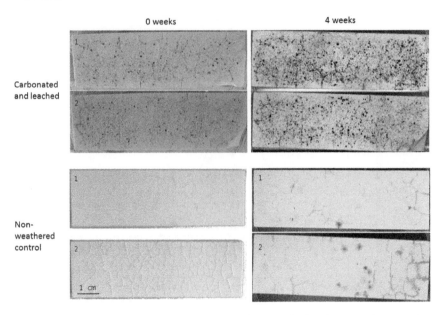

Figure 4.4 Effect of weathering (carbonation and leaching) on the development of *Conosporium unicinatum* on white Portland cement paste prisms [25].

4.4.2 Nutrients

The heterotrophic nature of fungi and the normal absence of organic compounds in concrete, means that these substances must usually come from external sources. However, one study has identified that mould-release agents used on the surface of concrete formwork can act as a source of energy and carbon for fungi [37].

Fungi also require sources of nitrogen, phosphorous, potassium, magnesium and sulphur. The presence of the first two nutrients in concrete is normally significantly limited, whilst potassium and magnesium may be present in slightly higher quantities. Sulphur is typically more abundant, since it is present as gypsum in Portland cement. This aspect is discussed in more detail in Section 3.4.3 of Chapter 3.

Regardless of availability, fungi possess the means of extracting these nutrients. Phosphorous is normally present in soil and rocks as insoluble minerals such as apatite. It has already been seen that one of the most common secondary metabolites of fungi is oxalic acid, and this is capable of releasing phosphate through the formation of the highly insoluble compound calcium oxalate [26]:

$$Ca_5(PO_4)_3OH + 5H_2C_2O_4 \rightarrow 5CaC_2O_4 + 3H_2PO_4 + H_2O.$$

In the same way, sulphur can be solubilized in the form of sulphate [27]:

$$CaSO_4 \cdot 2H_2O + H_2C_2O_4 \rightarrow CaC_2O_4 + H_2SO_4 + 2H_2O.$$

In the above example, gypsum is the source of sulphate, but this could also be a sulphate-containing cement hydration product such as monosulphate or gypsum.

Nutrients required in very small quantities by fungi include calcium, iron, copper, manganese, zinc and molybdenum. Whether these elements are present in concrete is more dependent on individual sources of constituent materials, but it is conceivable that all may be present in the relatively low concentrations required by fungi. It is worth noting that low concentrations of zinc, iron and particularly manganese appear to be stimuli for the production of larger quantities of citric acid (see Section 4.2) [6]. All of these metals form strong complexes with both oxalic acid and citric acid as shown in Tables 4.1 and 4.2 for Zn, Cu, Mn and Mo, and in Chapter 2 for Fe and Ca. Thus, fungi also have the means of releasing these micronutrients from soil, rock and, indeed, concrete.

However, this does not provide the whole picture—both citrate and oxalate salts of zinc, copper and manganese will be formed. Of these, the citrate salts are the more soluble ('slightly soluble', meaning 1–10 g/l), whilst the oxalate salts are considerably less soluble, although both sets of compounds are much more soluble than the oxides, which are likely to be the form present in soil and rock [31]. High concentrations of these metals are likely to be toxic to fungi, and so it would appear that the release of secondary metabolites provides a mechanism for controlling availability to within ranges which are beneficial for the organism [26]. Compounds formed by the reaction of molybdenum and oxalic and citric acids appear to be more soluble [32], although this element is much less abundant in the Earth's crust, and so harmful concentrations are unlikely to arise.

The issue of calcium is a still more interesting one. Unlike the other micronutrients, citric acid production is generally increased when calcium levels increase in solution [33]. Calcium will react with citric acid to produce calcium citrate. In calcareous rocks and concrete, the precipitation of calcium citrate within pores will lead to cracking and fragmentation. This process will be discussed in more detail later in this chapter, since it has serious implications for concrete durability. However, the fact that citric acid production is stimulated by the presence of calcium—and hence proximity to calcium-bearing minerals—suggests that this is another means by which fungi attempt to maintain levels of nutrients, since a cracked mineral surface is likely to yield more nutrients than one which is intact.

As well as producing oxalic acid, lichen produce a number of relatively large organic acid molecules which are members of the lichenic acids (Table 4.3). These acids are highly effective at chelating metal ions such as Al, Fe, Ca and Mg, and are likely to be produced as a means of breaking down minerals to release nutrients [35].

Table 4.1 Stability constants of complexes formed between citrate ions and fungal micronutrient elements.

Complex	Reaction	Stability Constant	Reference
Zn	$Zn^{2+} + C_6H_5O_7^{3-} \rightleftharpoons Zn(C_6H_5O_7)^-$	6.21	
	$Zn^{2+} + 2C_6H_5O_7^{3-} \rightleftharpoons Zn(C_6H_5O_7)_2^{4-}$	7.40	
	$Zn^{2+} + C_6H_5O_7^{3-} + H^+ \rightleftharpoons Zn(C_6H_6O_7)$	10.20	
	$Zn^{2+} + C_6H_5O_7^{3-} + 2H^+ \rightleftharpoons Zn(C_6H_7O_7)^+$	12.84	[28]
Cu	$Cu^{2+} + C_6H_5O_7^{3-} \rightleftharpoons Cu(C_6H_5O_7)^-$	7.57	
	$Cu^{2+} + 2C_6H_5O_7^{3-} \rightleftharpoons Cu(C_6H_5O_7)_2^{4-}$	8.90	
	$Cu^{2+} + C_6H_5O_7^{3-} + H^+ \rightleftharpoons Cu(C_6H_6O_7)$	10.87	
	$Cu^{2+} + C_6H_5O_7^{3-} + 2H^+ \rightleftharpoons Cu(C_6H_7O_7)^+$	13.23	
	$2Cu^{2+} + 2C_6H_5O_7^{3-} \rightleftharpoons 2Cu(C_6H_5O_7)_2^{2-}$	16.20	
Mn	$Mn^{2+} + C_6H_5O_7^{3-} \rightleftharpoons Mn(C_6H_5O_7)^-$	4.28	
	$Mn^{2+} + C_6H_5O_7^{3-} + H^+ \rightleftharpoons Mn(C_6H_6O_7)$	9.60	
Mo(VI)	$MoO_4^{2-} + C_6H_5O_7^{3-} + H^+ \rightleftharpoons MoO_4(C_6H_6O_7)^{4-}$	8.35	
	$MoO_4^{2-} + C_6H_5O_7^{3-} + 2H^+ \rightleftharpoons MoO_4(C_6H_7O_7)^{3-}$	15.00	
	$MoO_4^{2-} + C_6H_5O_7^{3-} + 3H^+ \rightleftharpoons MoO_4(C_6H_8O_7)^{2-}$	19.62	
	$MoO_4^{2-} + C_6H_5O_7^{3-} + 4H^+ \rightleftharpoons MoO_3(OH)(C_6H_8O_7)^-$	21.12	
	$2MoO_4^{2-} + 2C_6H_5O_7^{3-} + 4H^+ \rightleftharpoons Mo_2O_8(C_6H_7O_7)_2^{6-}$	31.02	
	$2MoO_4^{2-} + 2C_6H_5O_7^{3-} + 5H^+ \rightleftharpoons Mo_2O_8(C_6H_8O_7)(C_6H_7O_7)^{5-}$	35.86	
	$2MoO_4^{2-} + 2C_6H_5O_7^{3-} + 6H^+ \rightleftharpoons Mo_2O_8(C_6H_8O_7)_2^{4-}$	40.08	[29]
	$MoO_4^{2-} + 2C_6H_5O_7^{3-} + 4H^+ \rightleftharpoons MoO_4(C_6H_7O_7)_2^{4-}$	25.34	
	$MoO_4^{2-} + 2C_6H_5O_7^{3-} + 5H^+ \rightleftharpoons MoO_4(C_6H_8O_7)(C_6H_7O_7)^{3-}$	29.54	
	$MoO_4^{2-} + 2C_6H_5O_7^{3-} + 6H^+ \rightleftharpoons MoO_4(C_6H_8O_7)_2^{2-}$	33.34	
	$2MoO_4^{2-} + C_6H_5O_7^{3-} + 3H^+ \rightleftharpoons Mo_2O_8(C_6H_8O_7)^{4-}$	21.73	
	$2MoO_4^{2-} + C_6H_5O_7^{3-} + 4H^+ \rightleftharpoons Mo_2O_7(OH)(C_6H_8O_7)^{3-}$	26.90	
	$2MoO_4^{2-} + C_6H_5O_7^{3-} + 5H^+ \rightleftharpoons Mo_2O_7(OH)_2(C_6H_8O_7)^{2-}$	31.53	
	$4MoO_4^{2-} + 2C_6H_5O_7^{3-} + 9H^+ \rightleftharpoons Mo_4O_{13}(OH)_3(C_6H_8O_7)_2^{5-}$	60.76	
	$4MoO_4^{2-} + 2C_6H_5O_7^{3-} + 10H^+ \rightleftharpoons Mo_4O_{12}(OH)_4(C_6H_8O_7)_2^{4-}$	64.69	
	$4MoO_4^{2-} + 2C_6H_5O_7^{3-} + 11H^+ \rightleftharpoons Mo_4O_{11}(OH)_5(C_6H_8O_7)_2^{3-}$	77.45	

Table 4.2 Stability constants of complexes formed between oxalate ions and fungal micronutrient elements.

Complex	Reaction	Stability Constant	Reference
Zn	$Zn^{2+} + C_2O_4^{2-} \rightleftharpoons Zn(C_2O_4)$	3.88	
	$Zn^{2+} + 2C_2O_4^{2-} \rightleftharpoons Zn(C_2O_4)_2^{2-}$	6.40	
	$Zn^{2+} + C_2O_4^{2-} + H^+ \rightleftharpoons Zn(C_2HO_4)^+$	5.54	
	$Zn^{2+} + 2C_2O_4^{2-} + 2H^+ \rightleftharpoons Zn(C_2HO_4)^+$	10.76	[28]
Cu	$Cu^{2+} + C_2O_4^{2-} \rightleftharpoons Cu(C_2O_4)$	4.84	
	$Cu^{2+} + 2C_2O_4^{2-} \rightleftharpoons Cu(C_2O_4)_2^{2-}$	9.21	
	$Cu^{2+} + C_2O_4^{2-} + H^+ \rightleftharpoons Cu(C_2HO_4)^+$	6.31	
Mn	$Mn^{2+} + C_2O_4^{2-} \rightleftharpoons Mn(C_2O_4)$	9.98	
	$Mn^{2+} + 2C_2O_4^{2-} \rightleftharpoons Mn(C_2O_4)_2^{2-}$	16.57	
	$Mn^{2+} + 3C_2O_4^{2-} \rightleftharpoons Mn(C_2O_4)_3^{4-}$	18.42	
Mo	$MoO_4^{2-} + C_2O_4^{2-} + 2H^+ \rightleftharpoons MoO_4(C_2H_2O_4)^{2-}$	13.62	
	$2MoO_4^{2-} + 2C_2O_4^{2-} + 5H^+ \rightleftharpoons Mo_2O_7(OH)(C_2H_2O_4)_2^{3-}$	31.20	[30]
	$2MoO_4^{2-} + 2C_2O_4^{2-} + 6H^+ \rightleftharpoons Mo_2O_6(OH)2(C_2H_2O_4)_2^{2-}$	34.08	

4.4.3 Concrete as a physical habitat for fungi

The attachment of fungi to a surface, such as concrete, is usually assisted by the production of EPS to act as an adhesive. There is some disagreement in the literature with regards to the extent to which the roughness of a concrete surface encourages colonization by fungi. One study found that greater roughness was conducive to the growth of *Cladosporium sphaerospermum* [36]. Another—examining the growth of mixed communities of fungi—found little difference between surfaces of different roughness, although only three surface finishes were examined, two of which were similar [37].

As for the case of bacteria (Chapter 3), there is a strong correlation between the rate of colonization of a concrete surface and the water/cement ratio of the material [37]. This is shown in Figure 4.5, which plots coverage of mortar tiles by fungi versus this ratio. It is proposed by the authors of the study from which these results come that this is possibly the result of a higher surface area, higher moisture retention at the surface, or higher nutrient availability. The study was also the one which found no relationship between surface roughness and growth, which would appear to rule out

Table 4.3 A selection of lichenic acids and some species which produce them [34].

Acid	Lichen Species								
	Baeomyces roseus	*Cladonia furcata*	*Cladonia sylvatica*	*Caloplaca elegans*	*Parmelia conspersa*	*Parmelia stenophylla*	*Umbillicaria arctica*	*Evernia vulpina*	*Parmelia furfuracea*
Baeomycic acid	•								
Atranorin									•
Lecanoric acid									•
Olivetoric acid									•
Gyrophoric acid							•		

Table 4.3 contd. ...

...Table 4.3 contd.

Salazinic acid 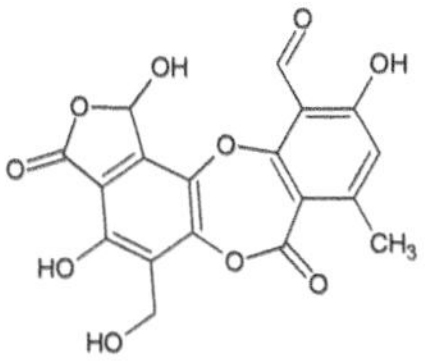 					•	•				
Fumarprotocetraric acid 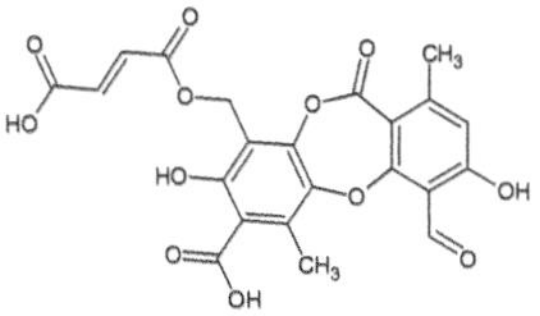 		•	•							
Lobaric acid 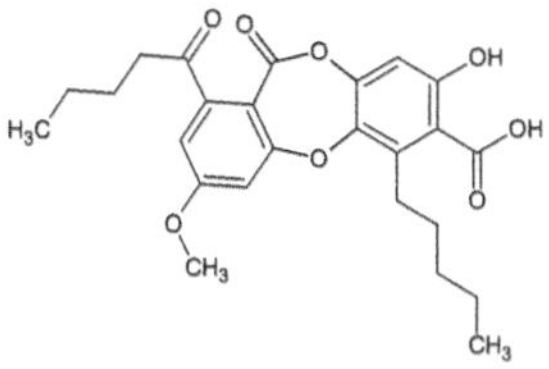 										
Physodic acid 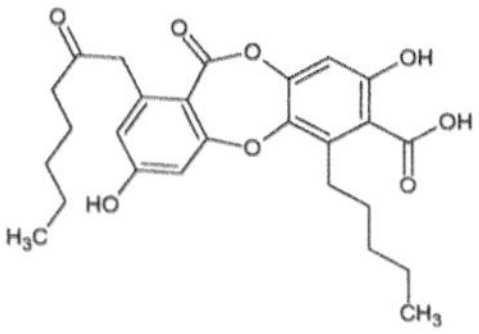 										•
Vulpinic acid 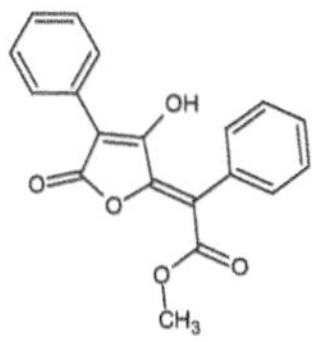 									•	
Ursolic acid 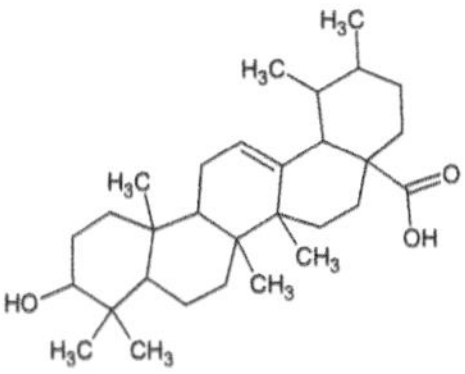 				•						

Table 4.3 contd. ...

...Table 4.3 contd.

Parietin				•					
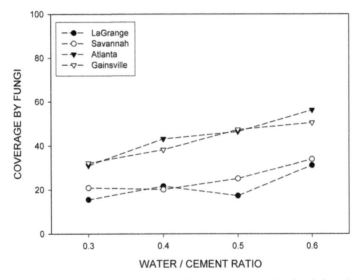 H₃C ...									
Usnic acid		•			•	•			

Figure 4.5 Fungal coverage of mortar tile specimens inoculated with fungal isolates derived from concrete surfaces from various locations in Georgia, USA. After inoculation, tiles were incubated in the dark with a source of nutrition (potato dextrose broth) for 7 days [37].

the first of these. Given that nutrients are also in relatively short supply in concrete itself, it is most probably the greater capacity of high water/cement ratio to retain moisture that is most important.

Once attached, hyphae begin to grow over the concrete surface. The hyphae are typically coated in a layer of EPS and exudates of secondary metabolites, thus forming a biofilm (see Figure 4.6). At locations beneath the biofilm, cracking and spalling of the concrete is sometimes observed. A scanning electron microscope image of a hypha is shown in Figure 4.7. Hyphae typically have diameters between 4 and 6 µm [40], although many of the fungi found growing on mineral surfaces have far narrower hypha—around 1 µm [41]. Nonetheless, this means that their penetration into the pores of pristine concrete is likely to be limited. Concrete can be damaged by fungi through both biochemical and biomechanical processes. Deterioration by biochemical means involves both the dissolution of the concrete by acidolysis, but also potentially the mechanical damage of the cement matrix through the precipitation of expansive salts (these processes are discussed in more detail in subsequent sections).

Biomechanical mechanisms principally involve the penetration of hyphae into mineral substrates. The mechanism by which this occurs is

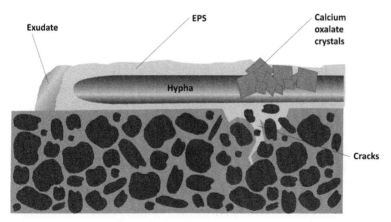

Figure 4.6 Interaction of a hypha growing on the surface of concrete [38].

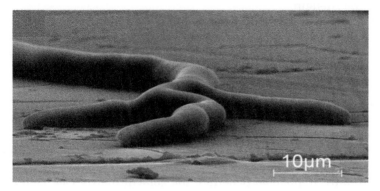

Figure 4.7 Cryo scanning electron microscope image of an *Aspergillus niger* hypha growing on a quartz surface [39].

through the widening of narrow openings through pressures exerted by the hyphae. These pressures are achieved through the development of turgor pressure of fluids within the hyphae through osmotic processes. Exerted pressures deriving from hyphae of fungi (albeit species which grow on plants) have been found to be around 0.05 MPa [42]. This is by no means enough to fracture concrete, which typically has tensile strengths two orders of magnitude higher. However, it is important to remember that the macro-scale properties of concrete will not be uniformly found throughout the material on a micro-scale. Moreover, the presence of melanin in the cell walls imparts rigidity which allows a much greater pressure to be exerted. One study, which used optical measurement techniques to determine force exerted by hyphae tips of a melanin-forming fungus recorded forces which were estimated to equate to an exerted pressure of 5.4 MPa [43]. It should be noted that the species investigated in this study (*Colletotrichum graminicola*) grows on plants, although most of the species known to inhabit concrete surfaces also produce melanin.

It has also been proposed that EPS exuded by fungi into fissures between mineral grains may act to physically damage mineral surfaces through swelling resulting from the absorption of water [44].

Regardless, it is likely in the case of fungi which reside on mineral surfaces that prior damage will have been done to the concrete via biochemical attack. One study which measured chemical and physical changes observed in Portland cement paste exposed to the culture fluid in which either *Aspergillus niger* or a sterile fungal strain were growing, used an experimental configuration which achieved this without physical contact between the fungus and the cement [45]. This approach ruled out any biophysical mechanisms of deterioration. In the case of *Aspergillus niger*, oxalic and gluconic acid were formed, whilst gluconic and malic acids were produced by the sterile fungal strain. The effect of exposure to the culture fluid leached calcium and increased porosity, leading to a decrease in flexural strength.

It should be noted that the presence of glucose in the culture would also have aided in the leaching of calcium, and the absence of a control exposed to the culture without fungi, means that the extent to which the organic acids play a role is not entirely clear. Indeed, the use of glucose in contact with cement and concrete in investigations of fungal deterioration is a common feature of such studies, and the damaging effect of this compound on its own must be borne in mind when interpreting results. Nonetheless, the development of porosity in this way is likely to lead to both more accessible routes into the cement matrix, and also a weaker matrix which offers more opportunity to be further broken by biomechanical means.

Certainly, hyphae *do* penetrate the cement matrix (see Figure 4.8) and there is strong evidence that the tips of hyphae can, under the right conditions, sense the route forward that offers the least resistance [39].

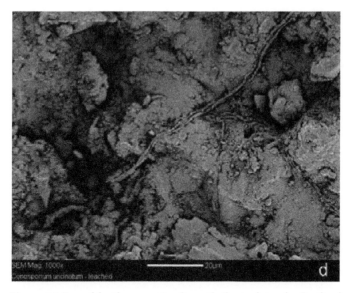

Figure 4.8 Fungal hyphae growing beneath the surface of a Portland cement paste [25].

4.5 Fungi on Concrete

Many studies have been conducted to characterise fungal species found on the surfaces of buildings, and it is likely that most of these species would be capable of growing on concrete. However, subsequent discussion will be limited to fungi which have been confirmed as being capable of growing on concrete surfaces. Table 4.4 lists fungi which have been successfully grown on concrete or cement paste surfaces in the laboratory, or identified on concrete in the field.

Another study has conducted more general characterisation of the biomass growing on the external walls of buildings in Europe and Latin America [46]. Whilst specific genera were not identified, fungi were found to make up the second most frequently encountered organism on buildings in Latin America (after cyanobacteria), but was the least common in Europe. The influence of the type of substrate was also examined: composite materials (in this context cement, mortar concrete and brick) tended to contain fungal life less frequently than they did algae or cyanobacteria, which was also the case for stone buildings. Painted surfaces were most frequently occupied by fungi.

Characterisation of fungi growing on the original concrete shield around the damaged nuclear reactor in Chernobyl in Ukraine has also been conducted [47]. These are listed separately from the other studies in Table 4.5. This is partly because it is not clear whether all samples were

Table 4.4 Fungi identified growing on concrete surfaces, or successfully grown in the laboratory.

Genera	Species	Fungi Division/ Class	Location	References
Alternaria	–	Ascomycota/ Dothideomycetes	Georgia, USA/São Paulo, Brazil	[48, 59]
	alternata		Laboratory	[25, 38, 54]
Aspergillus	–	Ascomycota/ Eurotiomycetes	São Paulo, Brazil	[59]
	glaucus		UK	[49]
	candidus		Moscow, Russia	[53]
	flavus		Moscow, Russia	[53]
	flavipes		Laboratory	[38, 54]
	fumigatus		Moscow, Russia	[53]
	niger		Laboratory/ Moscow, Russia	[50, 45, 38, 53, 54, 56, 57]
	ochraceus		Moscow, Russia	[53]
	repens		Laboratory	[38, 54]
	terreus		Moscow, Russia	[53]
	versicolor		Laboratory	[38, 51, 53, 54]
Aureobasidium	–	Ascomycota/ Dothideomycetes	São Paulo, Brazil	[59]
	pullulans		Moscow, Russia	[53]
Chaetophoma	–	Ascomycota/–	São Paulo, Brazil	[59]
Cladosporium	–		São Paulo, Brazil/ Rio Grande, Brazil/ Pirassununga, Brazil	[57a, 59]
	cladosporiodes		Georgia, USA	[48, 38, 53, 54]
	sphaerospermum		Laboratory/Daegu, South Korea	[36, 55, 56]
Coniosporium	*uncinatum*	Ascomycota/ Eurotiomycetes	Laboratory	[25]
Curvularia	–	Ascomycota/ Euascomycetes	São Paulo, Brazil	[59]

Table 4.4 contd. ...

...*Table 4.4 contd.*

Genera	Species	Fungi Division/Class	Location	References
Epicoccum	nigrum	Ascomycota/ Dothideomycetes	Georgia, USA	[48]
Exophiala	–	Ascomycota/ Chaetothyriomycetes	Laboratory	[25]
Fonsecaea	–	Ascomycota/ Eurotiomycetes	São Paulo, Brazil	[59]
Fusarium	–	Ascomycota/ Sordariomycetes	Laboratory	[52]
	–		Georgia, USA	[48]
Geotrichum	candidum	Ascomycota/ Saccharomycetes	Moscow, Russia	[53]
Helminthosporium	–	Ascomycota/ Dothideomycetes	São Paulo, Brazil	[59]
Monilinia	–	Ascomycota/ Ascomycetes	São Paulo, Brazil	[59]
Mucor	–	Zygomycota/ Mucormycotina	Georgia, USA	[48]
Nigrospora	–	Ascomycota/ Saccharomycetes	São Paulo, Brazil	[59]
Paecilomyces	lilacinus	Ascomycota/ Eurotiomycetes	Laboratory	[38, 54]
	varioti		Moscow, Russia	[53]
Penicillium	–	Ascomycota/ Eurotiomycetes	UK	[49]
	brevicompactum		Moscow, Russia	[53]
	chrysogenum		Laboratory	[51]
	expansum		Moscow, Russia	[53]
	funiculosum		Moscow, Russia	[53]
	islandicum		Moscow, Russia	[53]
	oxalicum		Georgia, USA	[48]
	purpurogenum		Moscow, Russia	[53]
	spinulosum		Moscow, Russia	[53]
	vermiculatum		Moscow, Russia	[53]
Pestalotia/ Pestalopsis	–	Ascomycota/ Sordariomycetes	São Paulo, Brazil	[59]

Table 4.4 contd. ...

...Table 4.4 contd.

Genera	Species	Fungi Division/ Class	Location	References
Pestalotiopsis	maculans	Ascomycota/ Sordariomycetes	Georgia, USA	[48]
Phoma	–	Ascomycota/ Dothideomycetes	São Paulo, Brazil	[59]
Rhizopus	oryzae	Zygomycota/ Mucormycotina	Moscow, Russia	[53]
	nigricans		Moscow, Russia	[53]
Scytalidium	–	Ascomycota/ Leotiomycetes	São Paulo, Brazil	[58]
Trichoderma	–	Ascomycota/ Pezizomycotina	São Paulo, Brazil	[59]
	asperellum		Georgia, USA	[48]
	viride		Moscow, Russia	[53]
Trichothecium	roseum	Ascomycota/ Pezizomycotina	Moscow, Russia	[53]

Table 4.5 Fungi identified growing on the concrete shield around the damaged reactor at Chernobyl [47].

Genera	Species	Genera	Species	Genera	Species
Acremonium	strictum	Chaetomium	globosum	Mucor	plumbeus
Alternaria	alternata	Chrysosporium	pannorum	Paecilomyces	variotii*
Aspergillus	flavus*	Cladosporium	cladosporioides	Penicillium	chrysogenum
	fresnii*		herbarum		citrinum*
	fumigatos		sphaerospermum		hirsutum
	niger		other species		hordei
	ochraceus*	Doratomyces	stemonitis		ingelheimense
	ustus*	Fusarium	merismoides	Phialophora	melinii*
Aureobasidium	pullulans		oxysporum	Stachybotrys	chartarum
	versicolor		solani	Sydowia	polyspora
Beauveria	bassiana*	Geotrichum	candidum*	Ulocladium	botrytis
Botyritis	cinerea		other species		

*Only found in locations where severe contamination with radioactive material had occurred (radiation levels = 40–220 mR/h).

all taken from concrete surfaces, and also because the researchers found different species in different proportions where levels of radiation were higher, presenting the possibility that the findings are not representative of a more conventional environment. Indeed, it would appear that some melanin-forming fungi are capable of radiotrophism—obtaining energy from ionizing radiation [60].

Nonetheless, many of the fungi identified from the Chernobyl site are also seen in Table 4.4.

In addition to the genera and species listed in Tables 4.4 and 4.5, a number of sterile mycelia growing on concrete surfaces have been identified in characterization studies [45, 47, 51]. Sterile mycelia *'mycelia sterilia'* are fungi that do not produce spores, meaning that characterising them in taxonomic terms is problematic.

Another fungus worthy of mention is *Serpula lacrymans*, or 'dry rot'. This is a fungus which lives and feeds on wood and can potentially cause severe damage to timber and timber-frame structures. However, it appears to benefit considerably from growth in close proximity to sources of calcium and iron. These sources can potentially include mortar, calcium silicate bricks, mineral insulation wools and concrete [61]. Where such a source exists, the fungus will grow over the surface of the calcium source, with possible growth of hyphae beneath the surface of the material [62]. The calcium is taken up by the hyphae, translocated through the mycelium and subsequently used in the formation of calcium oxalate through reaction with oxalic acid. As discussed previously, the acid is formed with the main objective of assisting in the breakdown of cellulose in the timber. However, it would appear that calcium oxalate formation is also used as a means of buffering the pH within a range which is optimal for the organism [63]. The white rot fungus *Resinicium bicolor* has also been found to use calcium translocation in a similar manner [64].

Whilst no evidence can be found in the literature of damage to concrete by *Serpula lacrymans*, the fact that calcium is removed from its source means that it is likely that damage of some magnitude occurs.

Table 4.6 lists species of lichen identified growing on concrete and mortar by researchers.

4.6 Chemical Deterioration from Fungal Activity

In the discussion of both the metabolism of fungi and nutrients available in concrete earlier in this chapter, it has become clear that oxalic and citric acid play an important role in the chemical interaction of concrete and fungi. Table 4.7 shows the results of a study examining the organic acids produced by filamentous fungi, having isolated the results which pertain to species found in Tables 4.4 and 4.5. The table identifies instances in which each acid was produced by the various strains of each species

Table 4.6 Lichen found growing on concrete and mortar surfaces.

Genera	Species	Growth Form	Reference	Genera	Species	Growth Form	Reference
Acarospora	cervina	Crustose	[67]	Lecanora	erysibe	Crustose	[65, 66]
	murorum	Crustose	[66]		turicensis	Crustose	[68]
	subcastanea	Crustose	[66]		albescens	Crustose	[65, 66, 69]
Aspicilia	Contorta ssp. hoffmanniana	Crustose	[68]		dispersa	Crustose	[66, 67, 70]
Caloplaca	sp.	–	[70]		muralis	Crustose	[66, 70]
	aurantia	Crustose	[68]		pruinosa	Crustose	[68]
	cinnabarina	Crustose	[66, 71]		umbrina	Crustose	[66]
	citrina	Crustose	[65, 66, 69, 71]	Lecidea	sp.	–	[70]
	erythrantha	Crustose	[65, 66]	Lepraria	sp.	Crustose	[68]
	erythrocarpa	Crustose	[68]		lesdainii	Crustose	[68]
	flavescens	Crustose	[66]		nivalis	Crustose	[66]
	holocarpa	Crustose	[65, 66, 67]	Leproplaca	xantholyta	Crustose	[68]
	lactea	Crustose	[68]	Opegrapha	calcarea	Crustose	[68]
	teicholyta	Crustose	[66, 67]	Phaeophyscia	chloantha	Foliose	[66]
	variabilis	Crustose	[68]	Physcia	undulate	Foliose	[66]

Genus	Species	Growth form	Ref.
Candelaria	velana	Crustose	[68]
	concolor	Foliose	[66]
Candelariella	aurella	Crustose	[66, 67, 70]
	vitellina	Crustose	[67]
Catapyrenium	squamulosum	Squamulose	[68]
Catillaria	lenticularis	Crustose	[66]
Clauzadea	immersa	Crustose	[68]
Collema	auriforme	Foliose	[68]
	crispum	Foliose	[68]
	tenax	Foliose	[68]
Dirina	massiliensis	Crustose	[68]
Dirinaria	picta	Foliose	[66]
Endocarpon	pusillum	Squamulose	[68]
Fulgensia	sp.	Crustose	[68]
Heterodermia	speciosa	Foliose	[66, 71]
Hyperphyscia	coralloides	Foliose	[66]
	syncolla	Foliose	[66, 69]
Protoblastenia	dubia	Foliose	[67]
	sp.	–	[70]
Pseudosagedia	linearis	Crustose	[68]
Punctelia	constantimontium	Foliose	[66]
	subpraesignis	Foliose	[66]
Pyxine	berteroana	Foliose	[66]
Rinodina	bischoffii	Crustose	[68, 70]
Sarcogyne	orbicularis	Crustose	[66]
	regularis	Crustose	[68]
Squamarina	concrescens	Squamulose	[68]
Staurothele	frustulenta	Crustose	[65, 66]
	monosporoides	Crustose	[66]
Teloschistes	chrysophthalmus	Fruticose	[66]
Toninia	aromatica	Crustose	[68]
Verrucaria	hochstetteri	Crustose	[68]
	macrostoma	Crustose	[68]
	muralis	Crustose	[68]
	nigrescens	Crustose	[67]

Table 4.6 contd. ...

...*Table 4.6 contd.*

Genera	Species	Growth Form	Reference	Genera	Species	Growth Form	Reference
	variabilis	Foliose	[66]	*Xanthoria*	*viridula*	Crustose	[68]
					sp.	–	[70]
	viridissima	Foliose	[66]		*candelaria*	Foliose	[66]
Lecania	*sp.*	Crustose	[66]		*fallax*	Foliose	[66, 69]
	cuprea	Crustose	[68]		*parietina*	Foliose	[65, 66]
	erysibe	Crustose	[65, 66]		*erysibe*	Crustose	[65, 66]
	turicensis	Crustose	[68]		*turicensis*	Crustose	[68]

Table 4.7 Instances of organic acid production produced by various strains of fungi of the genera *Aspergillus* [5].

	Acid														
Species	Acetic	Ascorbic	Butyric	Citric	Formic	Fumaric	Gluconic	Isobutyric	Itaconic	Lactic	Malic	Oxalic	Propionic	Succinic	Tartaric
Aspergillus flavus (2 strains)	1									1	2				
Aspergillus flavipes (1 strain)			1	1	1	1				1				1	
Aspergillus niger (14 strains)	1	4	7	11		1	8	3		1	10	14	1	9	9
Aspergillus terreus (1 strain)						1		1		1				1	

studied. It should be noted that more strains of *Aspergillus niger* were examined in the study than any of the other species. After oxalic and citric acids, malic, succinic, tartaric and gluconic and butyric acids are the most commonly encountered. The action of butyric acid on concrete has already been examined in Chapter 3. However, the remaining compounds are discussed below.

4.6.1 Oxalic acid

The action of oxalic acid on concrete is somewhat unusual, in that it can often have a protective effect. The reason for this can be seen in the solubility diagrams for oxalic acid in Chapter 2: in the case of Ca, Al and Fe, a solid phase is precipitated which persists to very low pH values. These solid phases are calcium oxalate monohydrate ($Ca(C_2O_4).H_2O$—whewellite), aluminium trioxalate tetrahydrate ($Al_2(C_2O_4)_3.4H_2O$) and iron (III) trioxalate pentahydrate ($Fe_2(C_2O_4)_3.5H_2O$).

We have seen in Chapter 3 that in the case of sulphuric acid attack the precipitation of salts resulting from reactions between acids and cement can be problematic if the molar volume of the precipitate is significantly greater than that of the hydration products it replaces. Whilst the molar volume of the iron and aluminium salts is not known, the molar volume of whewellite is 63.8 cm³/mol. Whilst this is larger than the molar volume of the portlandite that it will replace (32.9 cm³/mol) it is less than the molar volume of gypsum (74.5 cm³/mol). Thus, rather than cause expansion cracking in the manner of gypsum, a protective outer layer forms which prevent further ingress of acid. Thus, mass loss from cement paste and concrete specimens stored in solutions of oxalic acid is virtually zero [72].

The protective effect obtained by bringing oxalic acid in contact with a concrete surface is well-known, and there are a number of surface treatments available which are based on oxalic acid. This poses an important question: why can fungi cause damage to concrete seemingly, at least in part, through the release of an acid that would normally protect it.

There are, in fact, two forms of calcium oxalate which are commonly encountered in nature: whewellite and weddellite $(Ca(C_2O_4).2H_2O)$. In experiments where cement paste or concrete are exposed to relatively strong solutions of oxalic acid, whewellite is formed exclusively [73]. However, weddellite is encountered in other circumstances. Weddellite has a higher molar volume than whewellite: 79.2 cm^3/mol. This is also higher than that of gypsum, and it is therefore possible that its precipitation may have the potential to cause damage to concrete.

As discussed earlier, the formation of calcium oxalate by fungi occurs either in the hyphal walls or outside the hyphae. When produced within the organism, it would appear that lichen can control whether whewellite or weddellite are formed, with most species producing weddellite [74]. This ability to control calcium oxalate precipitation is likely to be possessed by fungi more generally. However, outside of the organism it must be presumed that fungi have much less control over calcium oxalate precipitation. Instead, the form precipitated is dependent on the ratio of calcium to oxalate, with a lower ratio favouring whewellite [68]. Presumably as a result of this, silicicolous lichens growing on calcareous surfaces tend to form weddellite, whereas they produce whewellite on siliceous surfaces [75].

Therefore, it is conceivable that damage can be done to concrete through the release of oxalic acid as long as the concentrations are sufficiently low that weddelite is formed. In support of this, crystals of calcium oxalate formed on concrete showing clear signs of expansion and deterioration under the action of *Aspergillus niger* were found to be composed of a mixture of the two calcium oxalate forms [38].

It should also be stressed that the protective effect of whewellite formation is observed when the entire surface of cement paste and concrete are exposed to a solution in which the concentration of oxalic acid is uniform. It is possible that the introduction of oxalic acid to localised points—as would be the case where fungi occupy the surface—may have a damaging effect. For instance, a small droplet of oxalic acid solution coming into contact with the surface of a large crystal of portlandite might create sufficient localized stress in the crystal lattice as a result of whewellite formation to cause it to fracture.

4.6.2 Citric acid

The nature of the interaction of citric acid with concrete is a process with the potential to cause significant damage. The reason for this has little to do with its strength as an acid, which is relatively low (see Chapter 2). Instead, the main process of deterioration is the precipitation of calcium citrate tetrahydrate ($Ca_3(C_6H_5O_7)_2.4H_2O$). There are, in fact, two different forms of this salt—the mineral earlandite [76] and a form which is obtained by precipitation through combining solutions of calcium and citric acid [77]. It is the second form which appears to form when citric acid comes into contact with cement [78].

Earlandite has a molar volume of 285 cm³/mol, and it is likely that the molar volume of the second form is similar. This is more than 8 times greater than portlandite (see Chapter 2). The effect—at least at higher concentrations—is an extremely rapid and progressive fragmentation at the concrete surface [73]. Figure 4.9 shows the development of deteriorated depth in Portland cement paste specimens with time for citric acid and oxalic acid. Results for acetic acid are also shown to provide comparison with an organic acid that largely causes deterioration by acidolyisis. The rate of deterioration in the case of citric acid is substantially greater than acetic acid.

Figure 4.10 shows mass loss from cement pastes of various types at two different citric acid concentrations, whilst Figure 4.11 shows micro-CT cross-sections through the specimens. In the case of the higher concentration of citric acid, the greatest resistance is displayed by the calcium sulphoaluminate cement, and the least by the PC/fly ash blend. This is also reflected in the CT scans, where the dramatic disintegration

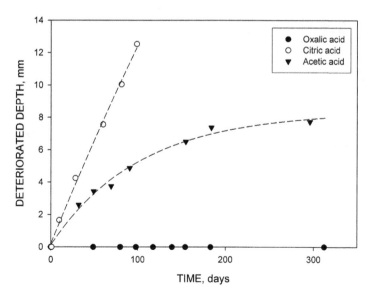

Figure 4.9 Development of degraded depth in cylindrical Portland cement paste specimens exposed to 0.1 M solutions of oxalic, citric and acetic acids [72].

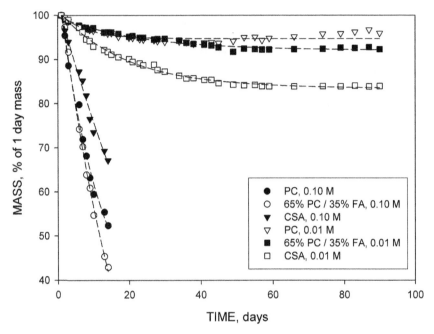

Figure 4.10 Loss of mass of cement paste specimens (water/cement ratio = 0.5) made from Portland cement (PC), a PC/FA blend, and a calcium sulphoaluminate cement exposed to 0.10 and 0.01 M solutions of citric acid [78] plus other data.

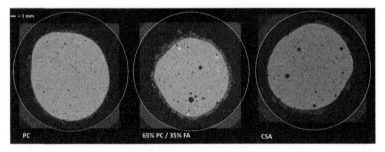

Figure 4.11 micro-CT scans showing cross-sections through cement paste specimens exposed to a 0.1 M solution of citric acid for 14 days [78].

of the cement is evident, with the fly ash specimen having undergone the greatest loss of material. The reason for this is most probably related to the relative abilities of the materials to resist fragmentation. All of the cement pastes had the same water/cement ratios, and so it is likely that the PC/FA blend would have the lowest strength, thus making it most susceptible to fragmentation. Since fragmentation will uncover the unaffected cement within, allowing it to be attacked, deterioration will progress at a faster rate.

At lower concentrations, different results are observed: the PC paste lost the least amount of mass, with the CSA specimen performing poorly. pH measurements of the exposure solutions indicate that the reason for the high performance of PC at lower concentrations is the result of the acid in the exposure solution being completely neutralized by the cement. This did not occur in the case of the PC/FA and CSA pastes. In the case of FA, this is simply because the paste contains less portlandite as a result of dilution by the fly ash and consumption during the pozzolanic reaction. However, in the case of CSA, neutralization had not yet occurred.

The process of neutralisation demonstrated using geochemical modelling is shown in Figure 4.12 for typical PC, 65% PC/35% FA blend and CSA compositions. It is evident that neutralization (indicated by a rapid increase in pH) occurs rapidly for the PC paste, takes longer for the FA blend and even longer for CSA. The neutralisation reaction where Portland cement is involved follows the reaction:

$$Ca(OH)_2 + 2H^+ \rightarrow Ca^{2+} + 2H_2O$$

Whereas calcium aluminate cements follow two reactions, the first of which is:

$$Ca_3Al_2O_9.6H_2O + 6H^+ \rightarrow 3Ca^{2+} + 2Al(OH)_3 + 6H_2O$$

Thus, an outer layer of $Al(OH)_3$ —gibbsite, albeit very poorly crystalline —develops at the surface of the cement. As more acid enters the cement matrix the pH drops further, at which point the $Al(OH)_2$ starts to decompose:

$$2Al(OH)_3 + 6H^+ \rightarrow 2Al^{3+} + 6H_2O$$

Where a calcium sulphoaluminate cement is involved the first reaction will be somewhat different, since it will involve aluminate hydration products such as ettringite:

$$Ca_6Al_2(SO_4)_3(OH)_{12}{\cdot}26H_2O + 6H^+ \rightarrow 6Ca^{2+} + 2Al(OH)_3 + 3SO_4^{2-} + 32H_2O$$

but the second reaction remains the same. The nature of the calcium aluminate cement reactions means that the material has a higher acid neutralization capacity than PC. However, the second reaction will only occur to any significant degree if the pH of the solution in contact with the $Al(OH)_3$ phase falls below around 5—this is because gibbsite remains effectively insoluble above this point, as seen in the solubility diagrams in Chapter 2.

The two-stage reaction of calcium aluminate cements means that the second neutralisation event is delayed. This delay can potentially be indefinite: where a weak acid or very dilute acid solution is present, the $Al(OH)_3$ layer at the surface may never dissolve, because the pH remains too high. Regardless, the delay gives the acidic species an opportunity to diffuse further into the material, causing further deterioration. This is

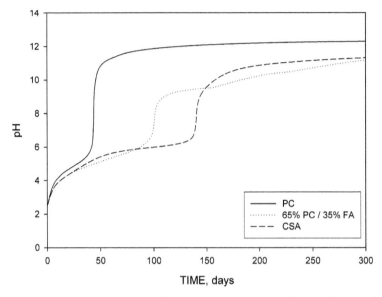

Figure 4.12 pH of 0.01 M citric acid solutions in contact with volumes of hardened cement paste with typical compositions for Portland cement (PC), a 65% PC/35% fly ash blend, and a calcium sulphoaluminate cement, calculated using geochemical modelling techniques. Solid:liquid ratio = 1:100.

what is occurring for the CSA cement paste exposed to the dilute solution in Figure 4.12.

The citrate ion is capable of forming relatively strong complexes with calcium, aluminium and iron. It is also capable of forming complexes with silicon (see Chapter 2). Within the context of concrete durability, this is of less concern, since fragmentation will typically remove material indiscriminately at a greater rate.

4.6.3 Tartaric acid

Tartaric acid is a relatively weak acid, but one which is capable of forming insoluble calcium tartrate on contact with cement. The influence of calcium tartrate precipitation on the integrity of cement is a mixture of positives and negatives.

Calcium tartrate is an expansive product, with a molar volume of 144 cm^3/mol. This effect is seen in Figure 4.13 which shows micro-CT scans obtained from a series of cement paste specimens exposed to a 0.1 M solution. The expansive nature of the product is evident from the images obtained from PC and PC/FA pastes, with a thick product having formed within the silica-gel layer of the specimens, which has been considerably disrupted, leading to partial delamination from the surface.

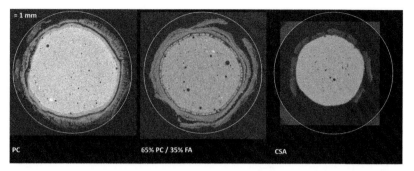

Figure 4.13 micro-CT scans showing cross-sections through cement paste specimens exposed to a 0.1 M solution of tartaric acid for 90 days [78].

However, a feature of this tartrate 'crust' is that, unlike calcium citrate, it does not fall away from the concrete, but remains attached, leading to an accumulation of material. This would appear to have something of a protective influence towards the concrete beneath it. Indeed, in the days when winemaking was commonly conducted against unlined concrete, the surface would be allowed to develop a crust of tartrate to limit further deterioration [79].

The effect of cement type on resistance to tartaric acid is also evident from Figure 4.13, with the PC/FA blend performing best, and calcium sulphoaluminate cement performing least well. The reason for this is possibly, in part, the absence of a tartrate layer.

Figure 4.14 plots the depth of the acid-deteriorated layer with time for a Portland cement paste exposed to a 0.10 M tartaric acid—without renewal of the solution. Results from a number of other acids including acetic acid are also included [73]. Figure 4.15 shows mass loss from the same set of experiments.

4.6.4 Malic acid

Malic acid also produces insoluble salts—calcium malate trihydrate and dihydrate (molar volumes 128.0 and 115.0 cm^3/g respectively). It is likely that calcium malate trihydrate is formed, since it is the least soluble of the two. Precipitation has a partly protective effect in the same manner as tartaric acid. However, comparing acetic acid and malic acid deterioration depth and mass loss results in Figures 4.14 and 4.15 it is evident that mass loss from Portland cement pastes is closer to that of acetic acid, whilst the deteriorated depth is much less, suggesting that fragmentation of the outer acid-deteriorated layer has occurred to some degree.

Figure 4.16 shows the results of geochemical modelling of malic acid attack on Portland cement. According to the model, only calcium malate

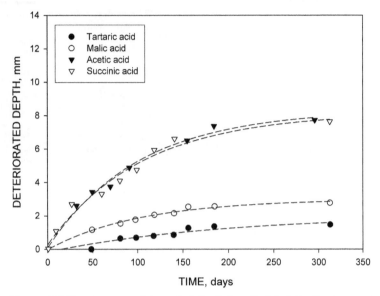

Figure 4.14 Development of degraded depth in cylindrical Portland cement paste specimens exposed to 0.1 M solutions of tartaric, malic, acetic and succinic acids [72, 73].

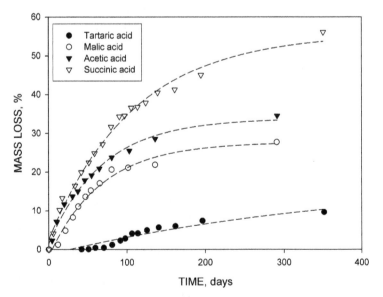

Figure 4.15 Mass loss from cylindrical Portland cement paste specimens exposed to 0.1 M solutions of tartaric, malic, acetic and succinic acids [72, 73].

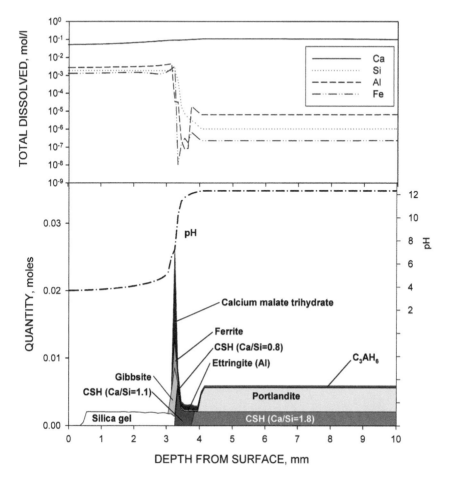

Figure 4.16 Concentrations of dissolved elements (top) and quantities of solid phases obtained from geochemical modelling of malic acid attack of hydrated Portland cement. Model conditions: acid concentration = 0.1 mol/l; volume of acid solution = 4 l; mass of cement 80 g; diffusion coefficient = 5×10^{-13} m^2/s.

trihydrate is precipitated as a very narrow band at the interface between the decalcified and partially decalcified zones.

4.6.5 Gluconic acid

There is a lack of experimental data relating to the deterioration of cement and concrete exposed to solutions of gluconic acid. However, some indication of its likely effect can be inferred from the data relating to the acid and its salts in Chapter 2. It is a somewhat stronger acid than the other compounds discussed in this section. Moreover, whilst it forms a salt with calcium (calcium gluconate monohydrate), it is more soluble and only likely to be formed where very high concentrations of the acid are present. For this reason, it is likely that deterioration will be primarily through acidolysis, which will be slightly more aggressive than for the other fungi-derived acids.

Figure 4.17 shows the results of geochemical modelling of attack by gluconic acid. Along with the absence of any salt precipitate, it is worth noting that ferrihydrite is absent, in contrast to the case of malic acid. This reflects the strong complexes that gluconic acid is capable of forming with iron (III) ions.

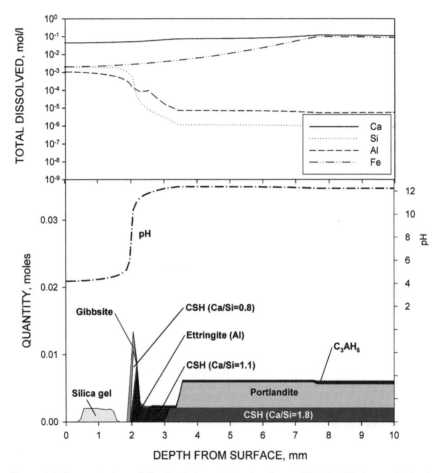

Figure 4.17 Concentrations of dissolved elements (top) and quantities of solid phases obtained from geochemical modelling of gluconic acid attack of hydrated Portland cement. Model conditions: acid concentration = 0.1 mol/l; volume of acid solution = 4 l; mass of cement 80 g; diffusion coefficient = 5×10^{-13} m^2/s.

4.6.6 Succinic acid

Succinic would appear to attack Portland cement at a rate which is comparable to acetic acid [67a]. This is not surprising, since it is only slightly weaker than this acid. Succinic acid reacts with calcium in cement paste to precipitate calcium succinate mono- and trihydrate. Unlike tartaric acid the formation of these compounds does not appear to have a protective effect, although it has been suggested that its formation is not detrimental [73]. However, by comparing Figures 4.14 and 4.15 it is evident that there is a discrepancy between the deteriorated depth results (which are comparable to acetic acid) and the mass loss results. This indicates that fragmentation of the outer acid-affected layer is most probably occurring, leading to greater rates of mass loss compared to acetic acid. Figure 4.18 shows the results of geochemical modelling of succinic acid attack on a PC paste.

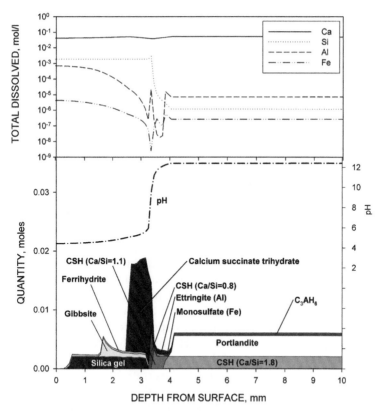

Figure 4.18 Concentrations of dissolved elements (top) and quantities of solid phases obtained from geochemical modelling of succinic acid attack of hydrated Portland cement. Model conditions: acid concentration = 0.1 mol/l; volume of acid solution = 4 l; mass of cement 80 g; diffusion coefficient = 5×10^{-13} m^2/s.

4.6.7 A broader consideration of deterioration from fungi-derived acids

Viewing the impact of acids commonly produced by fungi on concrete as a whole, it is evident that salt precipitation is a common feature, but also that the effect of precipitation is highly varied. In some instances the formation of salts leads to significant fragmentation of cement, whereas in other cases the effect is wholly or partly protective.

It is reasonable to assume that the molar volume of the salts plays an important role in influencing the effect that each acid has. The molar volume increases in the sequence oxalic < malic < succinic < tartaric < citric. The molar volume of calcium gluconate monohydrate is currently unknown. Whilst the calcium salts on the extreme end of this sequence correlate with the magnitude of damage observed, the other members do not.

Another factor which most probably plays a role in determining the nature of attack is the solubility of the salt. There are two reasons for this. Firstly, a lower solubility will cause a larger quantity of salt to be precipitated for a given acid concentration. Secondly, a lower solubility salt will be precipitated at closer proximity to the unaltered cement. This is significant, because if the salt has a high molar volume and is precipitated close to the intact cement, more damage will be done in comparison to a salt precipitated in a region of wholly decalcified cement.

Another interesting correlation between the magnitude of deterioration and the characteristics of the acids is shown in Figure 4.19. This plots logarithms of the first acid dissociation constants (pK_a) of the acids discussed above (minus gluconic acid) against the deteriorated depth observed in PC pastes at 100 days exposure to 0.1 M solutions. The reason for this correlation is not immediately obvious, and may be co-incidental. Indeed, it might be expected that an inverse correlation might be expected, since pK_a is a measure of acidic strength. Nonetheless, this relationship requires further investigation.

The ability of calcium aluminate cement to resist attack from the fungi-derived acids discussed is even less well understood. It has been seen that very different results are obtained when a calcium aluminate cement is exposed to tartaric and citric acid solutions, with the former displaying reduced resistance compared to PC, and the latter displaying superior resistance. Calcium aluminate cements exposed to oxalic acid develop the same type of protective oxalate layer. However, the behaviour of gluconic, succinic and malic acid is currently uncertain. Moreover, it is imprudent to attempt to estimate performance due to the erratic behaviour observed. For instance, geochemical modelling predicts that a protective calcium tartrate crust should be formed when tartaric acid comes into contact with calcium aluminate cement. Instead, a zeolite-like phase is formed, presumably as a result of the complexation of aluminium [78].

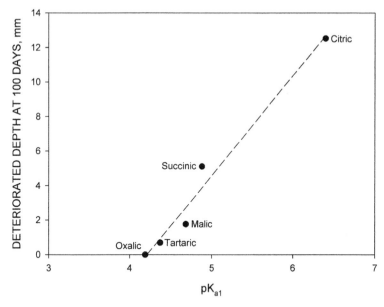

Figure 4.19 Deteriorated depth at 100 days of Portland cement pastes exposed to 0.1 M solutions of various organic acids produced by fungi versus the logarithm of each compound's first acid dissociation constant pK_{a1}. Deteriorated depth data from [72, 73].

4.7 Damage to Concrete from Fungi

This section will examine instances in the literature where fungi has been found to cause damage to hardened cement or concrete either in the laboratory or in the field. Instances where physical damage was observed are examined first, followed by a discussion of discolouration of concrete surfaces by fungi.

4.7.1 Physical damage

In reviewing research in the literature involving physical damage to concrete as a result of fungal activity, it is evident that *Aspergillus* and *Fusarium* species feature more frequently than other species. Whilst it can be safely concluded that these species are capable of causing deterioration of concrete, it is probably not sound to unquestionably conclude that they are the most common fungi which cause damage. The reason for this is that the discovery that such species damage concrete is likely to have stimulated further study using these species. However, it should be noted that studies where samples have been taken from deteriorating concrete in the field have all identified the two species as being present.

One of the earliest studies examining the influence of fungal activity on concrete durability looked at the effects of two species: *Aspergillus niger* and a sterile fungus [45]. Two experimental configurations were set up. In the first of these, the fungal strains were grown in a growth medium fluid which was continuously pumped onto the surface of cement paste samples, such that the fungi came into contact with the cement and grew on its surface. In the second, a similar process was followed, but a filter was placed between the fungal culture, meaning that only the culture fluid and any metabolites produced by the fungus came into contact with the cement.

The *Aspergillus niger* strain used in the experiments was found to produce gluconic and oxalic acid. The sterile fungus produced gluconic and malic acid. Where the cement came into contact with the organisms, there was substantial removal of calcium in the case of the sterile fungus, but no more than the control in the case of *Aspergillus niger*. One possible reason for this is evident from powder X-ray diffraction traces obtained from the cement paste specimens after exposure—calcium oxalate is present in the paste in contact with *Aspergillus niger* (although not commented on by the researchers). Thus, calcium was being retained in the cement paste as insoluble calcium oxalate.

Despite this retention of calcium, the porosity of cement specimens increased considerably in both cases, as illustrated in Figure 4.20, which shows the difference in pore size distributions relative to the control.

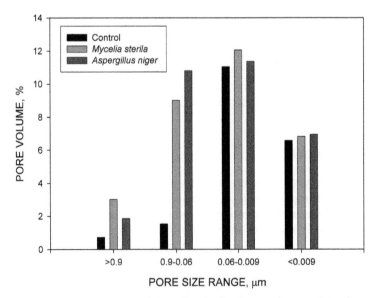

Figure 4.20 Changes in porosity in Portland cement pastes resulting from exposure to the culture fluid used to grow two fungi: a sterile fungi (*Mycelia sterila*) and *Aspergillus niger* [45].

The main change observed is an increase in the volume of porosity with diameters in the range 0.06–0.9 μm. This is attributed to the dissolution of crystals of Portlandite ($Ca(OH)_2$). There is also an increase in the volume of larger pores, which is attributed to the cracking of shrinkage-prone decalcified CSH gel at the cement surface after drying.

In addition there was a substantial loss in flexural strength resulting from contact with the fungi (Figure 4.21), although it should be noted that strength was not measured in control specimens and was still likely to be considerable due to the presence of glucose, which forms complexes with calcium and accelerates leaching.

Where the experiments were conducted such that the cement paste specimens only came into contact with the filtered culture fluid, similar results were obtained, indicating that the deterioration process was largely chemical. However, deterioration was greater in the case where *Aspergillus niger* grew on the cement itself.

Another laboratory experiment in which substantial damage was done to concrete by a fungal strain examined the growth of a *Fusarium* species on a section of concrete pipe sprayed with a culture medium acidified to pH 7 using hydrochloric acid (HCl) [52]. A considerably greater loss of calcium was estimated to be occurring from the pipe inoculated with *Fusarium* compared to the control, although this was based on measuring the quantity of HCl neutralised—an approach which employs assumptions which may not always be true. Nonetheless, the loss of mass from the concrete was

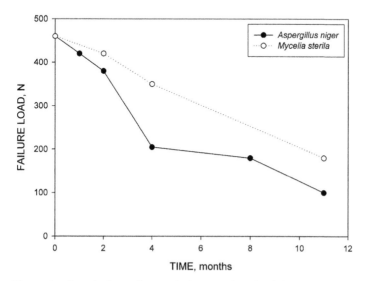

Figure 4.21 Loss in flexural strength (expressed as the failure load under three-point loading) resulting from growth of *Aspergillus niger* and a sterile fungus (*Mycelia sterila*) [45].

24% at 147 days exposure, after removal of loosely attached precipitates at the surface.

Experiments have been conducted examining the interaction of concrete chips in an agar growth medium inoculated with various fungal strains: *Alternia alternata, Aspergillus niger, Aspergillus versicolor, Cladosporium cladosporioides, Aspergillus repens, Aspergillus flavipes* and *Paecilomyces lilacinus* [38]. In almost all cases, discolouration of the cement occurred—the exception being in the case of *Aspergillus versicolor*. Additionally, a number of concrete specimens underwent expansion and cracking: *Aspergillus niger, Aspergillus versicolor, Cladosporium cladosporioides* and *Aspergillus flavipes*. Visual inspection indicated that *Aspergillus niger* caused the most substantial deterioration. SEM examination and powder X-ray diffraction analysis found crystals of calcium oxalate (in the form of both whewellite and weddelite) encrusting the concrete surface.

The researchers also conducted infrared spectroscopy and chemical analysis on droplets of exudate formed on *Aspergillus niger* hyphae growing on a concrete chip. They concluded that the exudate contained molecules with carboxylate groups which had formed complexes with calcium and silicon from the concrete.

A similar study examined the growth of *Alternaria alternata, Aspergillus flavipes, Aspergillus niger, Aspergillus versicolor, Aspergillus repens, Cladosporium cladosporioides, Paecilomyces lilacinus* on cement paste specimens [54]. In many cases discolouration, swelling and sometimes cracking were observed. The most substantial damage was observed in the case of *Aspergillus niger*, although deterioration was also observed in the case of *Alternaria alternata* and *Cladosporium cladosporioides*. Calcium oxalate formation (in the form of both whewellite and weddellite) was observed in the cases of all three of these fungi, plus *Aspergillus versicolor*. Cracking was observed for both *Aspergillus niger* and *Alternaria alternata*.

An evaluation of fungal inhabitation of concrete surfaces in two buildings in Moscow used for food processing—a large bakery plant and a meat processing plant—found a wide range of species (included in Table 4.4) and substantial deterioration [53]. This took the form of cracking and spalling of the concrete surface. Non-destructive testing of the colonised surfaces (probably using a rebound hammer, but not explicitly stated) found that the strength of the concrete had declined by up to 43% in the case of the bakery and 25% in the case of the meat processing plant. Another interesting feature of the concrete studied was that colonised surfaces had carbonated to a greater depth: 5–10 mm, compared to 2–3 mm in unaffected surfaces. This was attributed by the researchers as being the result of CO_2 production as a result of respiration of the fungus. This is wholly conceivable, but the presumably higher porosity of the deteriorated concrete is also likely to have played a role.

The most deteriorated area investigated was the basement of the bakery where there was a high quantity of airborne flour dust, which presumably promoted growth as a source of carbon and energy. *Aspergillus niger* was prevalent in this environment.

Deterioration of concrete by fungi was also identified in a UK study. In this case the fungi were growing below the waterline in a partially flooded cellar [49]. The fungi involved were sampled and isolated and identified as being *Aspergillus glaucus* and a *Penicillium* species. The isolated fungi were then placed in flasks with growth medium and small concrete specimens. A drop in pH of around 1.5 after 40 days was observed in the flask containing *Aspergillus glaucus*, which corresponded to a loss in mass from the concrete specimen of around 6%. Calcium was found to be the main element leached from the specimen during mass loss. The *Penicillium* fungus did not produce a drop in pH and mass loss was of a lesser magnitude.

The influence of two lichen species (*Acarospora cervina* and *Candelariella vitelline*) growing on asbestos cement roofs has been evaluated [65]. In the case of *Acarospora cervina*, the roof material on which it was growing was found to contain traces of whewellite, indicating oxalic acid production. This phase was not identified in the roof on which *Candelariella vitelline* was growing, which is known to produce pulvinic acid derivatives instead. Damage to the roofs was not quantified, but it was proposed that the lichen were altering the nature of the asbestos fibres, rendering them amorphous and less toxic. The study was conducted from the perspective of the health hazards associated with asbestos, and so this effect, plus the physical coverage of the roof by lichen which was presumed to prevent dispersal of asbestos fibres into the wider environment, was considered a favourable one.

One notable feature of the results of experiments investigating fungal deterioration of concrete is that, despite the presence of fungi known for the production of both oxalic and citric acid (such as *Aspergillus niger*), only calcium oxalate has been observed when the concrete is analysed using powder X-ray diffraction, and only in some cases. From the group of studies where this technique was used to analyse concrete or cement surfaces affected by fungi or lichen, only three positively identify calcium oxalate [38, 67, 54]. In all other cases it is absent [45, 68], although it is incorrectly identified in one instance.

The reason for this is not clear. It is certainly true that calcium oxalate is the less soluble of the two compounds, meaning that, where the two acids are present, calcium oxalate will precipitate in preference to calcium citrate. Thus, the absence of the citrate salt is conceivable. However, the absence of calcium oxalate is harder to explain. In some instances, the absence is simply because oxalic acid is not produced by the fungi. It is also possible that calcium oxalate is being dissolved by other acids. However, the salt remains insoluble even to very low pH, meaning that, if this were the case, the acid involved would need to be one which forms very strong complexes with calcium.

Oxalic acid is also decomposed to hydrogen peroxide and carbon dioxide when illuminated by sunlight in the presence of oxygen. Whilst direct decomposition of calcium oxalate does not appear to occur, dissolution of oxalate ions into water at a concrete surface would allow the gradual removal of oxalate. The reaction is significantly catalyzed by the presence of dissolved iron (III). It has been determined that under UV intensities of 0.65 m Einstein/m^2s, 1 µM dissolved iron (III) was able to decompose oxalic acid at a rate of 10 nM/s [80].

In at least one case, calcium oxalate was found to be absent at the fungi-inhabited surface of a cement paste specimen, despite analytical evidence that oxalic acid was being produced. It had previously been thought that once calcium oxalate was precipitated, fungi were subsequently unable to utilise the oxalate in any manner [81]. Recently, however, experiments conducted into calcium oxalate formation by a range of fungi found that, for some species, there was a subsequent disappearance of oxalate crystals [82]. It was proposed that the fungi were releasing enzymes such as decarboxylase or oxalate oxidase which were breaking down the oxalate to produce hydrogen peroxide (H_2O_2) which could be used in the oxidative decomposition of cellulose.

The fungi species found to be capable of this process were *Pleurotus tuberregium, Polyporus ciliates, Agaricus blazei, Pleurotus eryngii, Pleurotus citrinopileatus, Pycnoporus cinnabarinus* and *Trametes suaveolens*. It should be stressed that none of these species have been identified growing on concrete surfaces.

4.7.2 Discolouration of surfaces

In many cases, discolouration of concrete surfaces by fungi, rather than physical damage, may be the more important issue. In the case of fungi, discolouration occurs when the fungi produce pigments, the most common of which are the melanin compounds (see also Section 4.4.3). These are typically brown or black in colour. The majority of the species listed in Table 4.4 are known to be capable of producing melanins or melanin-like compounds.

The accretion of living and dead fungal cells, biofilm materials and pigments can lead to the formation of a black crust at the concrete surface. Moreover, biofilms have been likened to fly-paper, since their adhesive characteristics lead them to accumulate wind-borne dust and pollution particles [83].

In the case of lichen, whilst melanin may be produced by these organisms under certain circumstances [84], pigmentation will usually come from chlorophyll—used for photosynthesis by the photobiont, and imparting a green colour—and certain lichenic acids (see Section 4.4.2) which can impart a range of colours, including shades of yellow, brown, orange and red.

Whilst a discussion of aesthetic matters is beyond the scope of this book, it is probably reasonable to say that colour changes imparted to the surface of buildings and other components of the built environment by the colonisation of a surface by lichen (Figure 4.22) is often considerably less disagreeable to the eye in comparison to the dark crusts produced by many fungi. As is hopefully evident from Section 4.7.1, there would appear to be a lack of reported evidence of damage to concrete from lichen. Moreover, there is some evidence that lichen layers, in some instances, act to physically

Figure 4.22 Lichen growth on a concrete surface.

protect surfaces [67, 85]. With this in mind, the decision of whether lichen needs to be removed on aesthetic grounds needs to be made on the basis of the nature of the building, its surroundings and, potentially, the opinions of its users.

Nonetheless, there are functional reasons why both fungi and lichen may need to be removed. One of these is that, in wet conditions, colonised surfaces may present a slip hazard. Moreover, where concrete surfaces are adjacent to timber components in a structure, removal may be prudent to avoid colonisation and deterioration of the wood by fungi.

4.7.3 Corrosion of steel

Fungi have been found to be capable of inducing corrosion in steel tendons in post-tensioned concrete. The technique utilizes tendons held within polymer sheaths which are placed into tension after construction of a structural element is complete. To allow the tendon to move freely during tensioning, a lubricating grease is present within the sheath. A study investigating the possibility of fungal corrosion isolated species of *Fusarium*, *Penicillium* and *Hormoconis* from grease in cables taken from an existing structure [87].

The isolated organisms were then used to inoculate sheathed tendons, which were then sealed and held at a temperature of 23°C, alongside control specimens which had only had distilled water introduced into them. The inoculated specimens displayed localized corrosion. Most of the controls displayed no corrosion, with the exception of a sheath that later tested positive for *Fusarium*.

Infra-red spectrometry was employed to examine chemical changes in the grease, and identified the development of organic acids with carboxylate and hydroxy groups, leading the researchers to conclude that corrosion was induced by fungi oxidizing the lubricant (a metal soap hydrocarbon grease) as a source of energy.

4.8 Limiting Fungal Deterioration

Measures for limiting fungal deterioration can be achieved in three ways. Firstly, the composition of the concrete can be formulated such as to either limit the extent to which the concrete surface permits colonization, or such that resistance to attack is enhanced. Secondly, the concrete surface may be treated to resist colonization. Thirdly, where the surface is already colonized, cleaning and maintenance may be conducted. These three approaches are discussed below. The approach taken throughout is to discuss established techniques before discussing more developmental approaches.

4.8.1 Concrete composition

It is uncertain as to whether cement type plays any role in encouraging or discouraging colonisation and growth by fungi. A study which examined the development of fungi on mortar surfaces examined the influence of including ground granulated blastfurnace slag (up to 50%), fly ash (to 25%), silica fume (to 15%) and metakaolin (8%) [37]. Little difference was found in the rates of growth.

Another study found that higher levels of GGBS in combination with PC in cement pastes did limit fungal growth (in this case a mixture of *Aspergillus niger, Chaetomium globosum, Penicillium funiculosum* and *Gliocladium virens*) [80]. Levels of GGBS up to 80% by mass were used, and levels of 65% and over appeared to limit growth entirely after a period of 9 months in contact with a growth medium. Leach tests were carried out on the slag and no organic or inorganic substances likely to have a detrimental influence on fungal growth were detected. The researchers proposed that the high pH of the GGBS was responsible, but this seems unlikely, since PC will have a higher pH.

Further analysis of the data from this study presents another possibility: the pastes with high slag contents contained lower levels of gypsum. This correlation is shown in Figure 4.23. Given the importance of sulphur as a

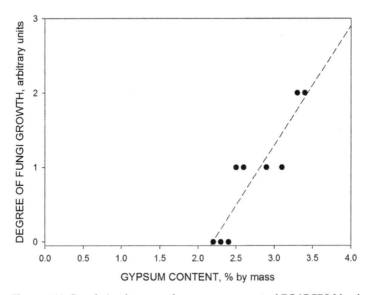

Figure 4.23 Correlation between the gypsum content of PC/GGBS blend cement pastes and the degree of fungal growth. Fungal growth is quantified using a scale employed by the Slovakian standard test for measuring resistance to mould growth on construction materials. 0 = no growth; 1 = negligible growth; 2 = 25% coverage [88].

nutrient (see Section 4.4.2) the limited sulphate levels in the cement may be the reason for the absence of fungi.

Regardless of its influence of on fungal growth rates, cement type will also influence rates of deterioration. There is incomplete understanding of this area with respect to the acids typically produced by fungi, and performance is dependent on the type of acids. However, it would appear that, in most cases PC or calcium aluminate cements offer a slight advantage. In the case of tartaric acid, PC/fly ash blends, and possibly PC slag blends may be more beneficial. A feature of many of the acids formed by fungi is that they form an insoluble and often expansive salt with calcium. Moreover, biophysical processes operate by generating stresses within concrete which lead to fracture. There is, therefore, a very strong argument, where fungal attack is a possibility, for employing concrete with a low water—cement ratio, and a hence high strength to resist the stresses associated with expansion. This has the secondary benefit of limiting rates of colonisation and growth through reduced water absorption (see Section 4.4.3).

The mechanical aspects of attack from fungi means that the choice of aggregate used will have a lesser influence on performance in comparison to modes of deterioration such as bacterial attack, where acidolysis can be checked by the neutralising capacity of calcareous aggregate. Indeed, calcareous aggregate will presumably also form expansive salts in contact with fungi-derived acids.

Another approach to limiting fungal growth employs titanium dioxide (TiO_2) powder. The use of TiO_2 in concrete exposed to the atmosphere as a means of maintaining a clean surface and reducing atmospheric pollution has been a developing technology over the past 20 years. TiO_2 acts as a photocatalyst: when illuminated with sunlight the surface of the compound reacts with oxygen and water in the atmosphere to produce oxidising and reducing species which are subsequently able to react with and break down various airborne pollutants including oxides of nitrogen (NO_x), sulphur dioxide (SO_2) and various volatile organic compounds (VOCs) [89]. The anatase form of TiO_2 is most effective. These reactions also act on non-volatile organic substances which have become attached to the concrete surface, ultimately leading to their degradation to an extent which allows them to be easily washed away from the surface by rainwater.

The effectiveness of this so-called 'self-cleaning' characteristic of concrete containing TiO_2 with regards to preventing the colonisation and growth of fungi on concrete found the approach to be successful [37]. A series of mortar tiles were prepared containing a small quantity (unspecified) of anatase nanoparticles in the cement. The tiles were inoculated, along with mortar tiles made with the same cement without TiO_2, with various fungal strains and incubated for a week in an environment where a nutrient solution was sprinkled periodically onto their surface. The tiles were lit for 6 hours with near-UV radiation followed by 6 hours of darkness. After incubation, the tiles with TiO$_2$ were found to be considerably more resistant to fungal colonisation than the controls.

Whilst TiO$_2$ particles in concrete appear to be effective against fungal colonization, long-term performance is currently not well-understood. Certainly it appears that where levels of surface contamination by dirt are high, the material does not perform as well [90]. Moreover, where surfaces are horizontal, and hence the self-cleaning process is not assisted by gravity, accumulation of dirt—and presumably fungus—is still possible.

It is possible to add fungicidal admixtures at the mixing stage of concrete production. There are a wide variety of substances with fungicidal effects. These include compounds containing copper, zinc or boron, disinfectants (such as sodium hypochlorite), quaternary ammonium compounds and various organic compounds which are toxic to fungi. However, when used as an admixture it is also necessary that a fungicide does not have a detrimental effect on concrete properties.

Much of the guidance regarding anti-fungal admixtures identifies polyhalogenated phenols and copper compounds as suitable candidates, but this is somewhat outdated [91].

Research into various potential formulations for copper-based antifungal admixtures examined a wide range of soluble candidate compounds [92]. The study measured growth of *Aspergilus niger* and *Penicillium viridis* on the surfaces of concrete specimens made from concrete which had various concentrations of the compounds added. Of the compounds evaluated, copper arsenite and copper acetoarsenite were found to be most effective, although additions of at least 5% by mass of cement were required. The use of copper based admixtures, however, has side-effects. The presence of copper will normally retard the hydration reactions of Portland cement [93], and it was found that the compressive strength of the concrete mix containing 10% copper acetoarsenite was around 80% less than the control. Setting time was, in fact, accelerated. Additionally, copper-based admixtures will normally impart an—often dramatic—colour change to concrete.

The main concern with such admixtures however, is the presence of arsenic which is hazardous to human health and a possible carcinogen. As a result, these compounds are now restricted in many countries. Other copper-based compounds may potentially suitable, but the retardation of cement hydration is far from ideal.

Polyhalogenated phenols have also been demonstrated as being effective against mould growth when used as a concrete admixture [94]. Again, there are concerns regarding the human health and environmental hazards presented by many of these compounds, and many substances are now strictly regulated. Indeed, the main reason for exercising caution before deciding upon the use of biocidal admixtures in concrete is the potential for releasing ecotoxic substances into the wider environment.

The search for less environmentally harmful antifungal admixtures has been challenged by the need to identify substances that do not detrimentally influence the strength development of Portland cement. For instance, one study found that both sodium hypochlorite and a mixture of chlorhexidine and cetrimide reduced concrete strength considerably and increased porosity [95].

The use of two other organic fungicidal agents as admixtures has been investigated more recently [96]. These were nitrofuran and an isothiazoline/carbamate mixture. The latter of these displayed an ability to limit growth of *Aspergillus niger* when included as a constituent in mortar. Moreover, additions of this formulation had negligible influence over both compressive and flexural strength development, even up to the largest dosage (5% by mass of the cement).

In the place of these compounds, a more recent development in fungicidal admixture technology has been the introduction of formulations which employ a so-called 'electro-physical' mechanism [97]. The compounds involved are silane compounds which, like water-repellent admixtures, attach to the surface of hydration products in the cement matrix. Joined to the silane group is a polymer chain component which possesses a positively charged group which electrostatically attracts, and then ruptures, the outer cell membranes of micro-organisms. The precise nature of this charged group is unclear, although patents exist for similar products containing molecules comprising silane groups joined to quaternary ammonium groups [98]. This approach clearly has benefits: by anchoring the molecules to surfaces, the admixture should, in theory, last indefinitely. Additionally, anchoring presumably limits the release of biocide into the wider environment.

Patents also exists for cement containing the mineral colemanite $(Ca[B_3O_4(OH)_3].H_2O)$ [99]. This mineral is known to have anti-fungal characteristics as a result of its boron content [100], although its effectiveness when used as an admixture in concrete is undocumented.

Many concrete admixtures also contain fungicidal/bactericidal additives to aid in preserving the admixture itself prior to use. The extent to which such substances contribute towards protecting hardened concrete during service is unknown, but is presumably small.

Another approach under development is the possible use of fungicidal micro-capsules [101]. These consist of granules of between 300 and 1000 μm. The outside of the capsule consists of a thin polyurethane membrane. Inside is a pellet consisting of a mixture of a fungicide of relatively low human toxicity (D-limonene, a naturally-occurring biocide found in the skin of citrus fruits) and particles of zeolites. The zeolite particles are present to provide structure and mechanical strength to the pellet. The micro-capsules are included in concrete and mortar as a partial replacement of the aggregate.

The principle behind the micro-capsules is that, as time passes, the membrane will become ruptured leading to a slow release of the biocide. The presence of the capsules reduced strength (10% at a level of 15% by mass of cement) and increased drying shrinkage. However, it was effective in limiting fungal growth on concrete surfaces on which fungi grew substantially in the absence of the capsules.

A similar approach has been adopted where the delivery mechanism is through polymer fibres impregnated with biocides [102, 103, 104]. The biocide is introduced during the fibre extrusion process. In the case of the impregnation of polypropylene fibres with a commercial biocide based on 2,4-dichlorophenol, inclusion of fibres at a level of 0.025% by mass of concrete was found to limit mould growth in laboratory experiments [102].

4.8.2 Protection after construction

After construction, paints may also be used to place a protective barrier between the concrete and the damaging effects of fungal activity. It should be noted that the presence of a coating of paint will not necessarily prevent the establishment and growth of fungus, although it does affect the types of species involved to some extent [48]. Silicate paints, however, do possess the potential to limit fungal colonisation [105]. This is for two reasons. Firstly, the paint tends to be alkaline, which presents a hostile environment for fungi (see Section 4.4.1). Secondly, silicate paints allow the concrete to 'breathe'—water vapour is not prevented from leaving the concrete pores, thus limiting the extent to which accumulation of moisture beneath paint surfaces might otherwise encourage fungal growth.

Treatment with hydrophobic impregnants, such as silane compounds, is a well-established approach for increasing the water repellency of concrete surfaces. These compounds have also been used in the preservation of stone to prevent fungal colonisation [106]. The main reason for their effectiveness is their ability to limit the extent to which water can accumulate at pores in the concrete, which is a key requirement for fungi (see Section 4.4.3). It is also likely that the presence of silanes at the surface prevents the ingress of acids. Whilst there is little by way of scientific evidence demonstrating whether this approach is appropriate for controlling fungal colonisation of concrete structures, it seems reasonable to assume that it would be. Indeed, the presence of fungi and algae is proposed as evidence—during visual inspection of concrete structures—that a surface has been inadequately treated with silanes [107].

Some concerns have been raised with regards to the longevity of such treatments [108]. However, when used correctly, a service life of around 15 years can be expected for most commercially-available products [109]. It should, however, be stressed that silane treatments *limit* the extent to which water can penetrate pores, and cannot entirely prevent it. Sufficient pressures are able to overcome the water repellency. These pressures need not be high: a wind-blown raindrop travelling at a reasonably high horizontal velocity might quite easily generate sufficient pressure to enter a concrete pore despite treatment with a silane or similar compound [109].

A traditional approach used on masonry, but not commonly encountered in the field of concrete construction, is the use of strips of copper flashing on structures [110]. The strips are typically set into mortar joints. As rainwater runs over the metal strip a small quantity of copper is dissolved, which has a similar effect to a copper biocide treatment. Whilst this approach is apparently highly effective, it will usually leads to a green discolouration of the masonry surface.

4.8.3 Cleaning and maintenance

The growth of fungus and lichen on concrete surfaces can, of course, be checked by periodic cleaning. This can be achieved through brushing with wire brushes or similar abrasive methods, scraping, and the use of high-power water jets. This approach may not be appropriate where abrasion of the concrete surface is to be avoided, such as where a high quality surface finish must be preserved. The case of the cleaning of asbestos cement roofs has been identified as another instance where abrasion of the surface must not occur, since it has the potential to release hazardous asbestos fibres [111]. In such cases, the use of fungicidal treatments may be a possibility.

A study examining the possibility of treating lichen-colonized asbestos cement roofs evaluated a wide range of anti-fungal treatments in terms of both their ability to remove lichen and their persistence in preventing re-colonization of the surfaces [112]. The results of this evaluation are shown in Figure 4.24. The most effective treatments were copper and boron compounds, plus a commercial formulation based on methyl 1-(butylcarbomoyl)-2-benzimidazolecarbamate and 5,6-dihydro-2-methyl-N-phenyl-1,4-oxathiin-3-carboxamide. Many other treatments had very little influence on fungal growth, and potassium permanganate actually encouraged growth. Given the role of both potassium and manganese as nutrients, this is not perhaps surprising.

One proposed approach to controlling fungal colonisation under development is the use of other organisms which are capable of inhibiting fungal growth. One study has evaluated the effectiveness of the bacteria *Bacillus aryabhattai* for this purpose, and found it capable of inhibiting growth of *Cladosporium sphaerospermum* [55].

The same research group has subsequently broadened its search for antifungal bacteria. This investigation concluded that *Bacillus aryabhattai* (and another bacteria—*Bacillus thuringiensis*) was less effective at preventing fungal growth on cement pastes than *Arthrobacter nicotianae*, *Bacillus thuringiensis* and *Stenotrophomonas maltophilia* [56]. Growth of

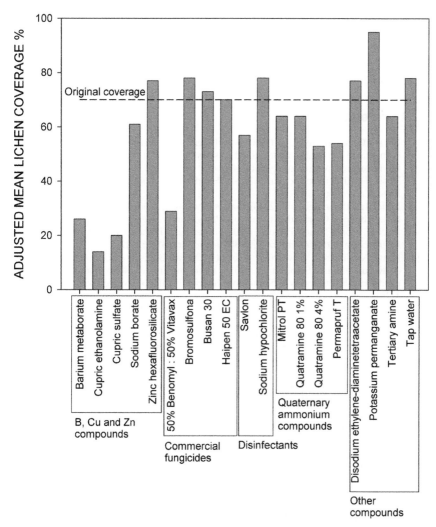

Figure 4.24 Lichen coverage of asbestos cement roof panels after treatment with potentially fungicidal additives. Panels initially had c 70% lichen coverage. After treatment, panels were left outside for a period of 83 months prior to evaluation of coverage [112].

both *Aspergillus niger* and *Cladosporium sphaerospermum* species were investigated. Further research has also identified the effectiveness of *Paenibacillus polymyxa* E681 against *Aspergillus niger* [57]. In all of these cases, the bacteria were selected not only on their ability to control fungal growth, but also in their ability to form calcium carbonate minerals to repair cracks in concrete. The research represents a relatively early stage in understanding, and it is not yet clear how these bacteria would be best deployed in the built environment. However, research into technologies which exploit calcite forming bacteria for concrete repair have devised a number of possible means by which such bacteria can be introduced into fresh concrete with the possibility of subsequent release where and when damage occurs.

4.9 References

[1]　Buck JW and Andrews JH (1999) Attachment of the yeast *Rhodosporidium toruloides* is mediated by adhesives localized at sites of bud cell development. Applied and Environmental Microbiology **65(2):** 465–471.

[2]　Nash TH (2010) Introduction. *In:* Nash TH (ed.). Lichen Biology. Cambridge University Press, Cambridge, UK.

[3]　Büdel B and Scheidegger C (2010) Thallus morphology and anatomy. *In:* Nash TH (ed.). Lichen Biology. Cambridge University Press, Cambridge, UK.

[4]　Lisci M, Monte M and Pacini E (2003) Lichens and higher plants on stone: a review. International Biodeterioration and Biodegradation **51(1):** 1–17.

[5]　Liaud N, Giniés C, Navarro D, Fabre N, Crapart S, Herpoël-Gimbert I, Levasseur A, Raouche S and Sigoillot J-C (2014) Exploring fungal biodiversity: organic acid production by 66 strains of filamentous fungi. Fungal Biology and Biotechnology **1(1):** 1–10.

[6]　Jennings DH and Lysek G (1996) Fungal Biology: Understanding the Fungal Lifestyle. Bios, Oxford, UK.

[7]　Gadd GM (1999) Fungal production of citric and oxalic acid: importance in metal speciation, physiology and biogeochemical processes. Advances in Microbial Physiology **41:** 47–92.

[8]　Dutton MV and Evans CS (1996) Oxalate production by fungi: its role in pathogenicity and ecology in the soil environment. Canadian Journal of Microbiology **42(9):** 881–895.

[9]　Gharieb MM and Gadd GM (1999) Influence of nitrogen source on the solubilization of natural gypsum ($CaSO_4.2H_2O$) and the formation of calcium oxalate by different oxalic and citric acid-producing fungi. Mycological Research **103(4):** 473–481.

[10]　Takao S (1965) Organic acid production by Basidiomycetes. I. Screening of acid-producing strains. Applied Microbiology **13(5):** 732–737.

[11]　Horner HT, Tiffany LH and Cody AM (1983) Formation of calcium oxalate crystals associated with apothecia of the discomycete *Dasyscypha capitate*. Mycologia **75(3):** 423–435.

[12]　Fink S (1991) Unusual patterns in the distribution of calcium oxalate in spruce needles and their possible relationships to the impact of pollutants. The New Phytologist **119(1):** 41–51.

[13]　Hang YD, Splittstoesser DF, Woodams EE and Sherman RM (1977) Citric acid fermentation of brewery waste. Journal of Food Science **42(2):** 383–384.

[14]　Pintado J, Torrado A, González MP and Murado MA (1998) Optimization of nutrient concentration for citric acid production by solid-state culture of *Aspergillus niger* on polyurethane foams. Enzyme and Microbial Technology **23(1-2):** 149–156.

[15]　Clark DS, Ito K and Horitsu H (1966) Effect of manganese and other heavy metals on submerged citric acid fermentation of molasses. Biotechnology and Bioengineering **8(4):** 465–471.

[16]　Vandenberghe LPS, Soccol CR, Pandey A and Lebeault J-M (1999) Microbial production of citric acid. Brazilian Archives of Biology and Technology **42(3):** 263–276.

[17]　Kubicke CP, Zehentgruber O, El-Kalak H and Roehr M (1980) Regulation of citric acid production by oxygen; effects of dissolved oxygen tension on adenylate levels

and respiration in *Aspergillus niger*. European Journal of Applied Microbiology and Biotechnology **9(2):** 101–116.

[18] Gorbushina AA (2006) Fungal activities in subaerial rock-inhabiting microbial communities. *In*: Gadd GM (ed.). Fungi in Biogeochemical Cycles. Cambridge University Press, Cambridge, UK.

[19] Honegger R (2001) The symbiotic phenotype of lichen-forming ascomycetes. pp. 165–188. *In*: Hock B (ed.). The Mycota, Volume 9: Fungal Associations. Springer, Berlin.

[20] Mason CF (1977) Decomposition. Arnold, London, UK.

[21] Rousk J, Brookes PC and Bååth E (2009) Contrasting soil pH effects on fungal and bacterial growth suggest functional redundancy in carbon mineralization. Applied and Environmental Microbiology **75(6):** 1589–1596.

[22] Fries L (1956) Studies in the physiology of Coprinus. II. Influence of pH, metal factors and temperature. Svensk Botanisk Tidskrift **50:** 47–96.

[23] El-Abyad MSH and Webster J (1968) Studies on pyrophilous discomycetes. I. Comparative physiological studies. Transactions of the British Mycological Society **51(3-4):** 353–367.

[24] Yamanaka T (2003) The effect of pH on the growth of saprotrophic and ectomycorrhizal ammonia fungi *in vitro*. Mycologia **95(4):** 584–589.

[25] Wiktor V, Grosseau P, Guyonnet R, Garcia-Diaz E and Lors C (2010) Accelerated weathering of cementitious matrix for the development of an accelerated laboratory test of biodeterioration. Materials and Structures **44(3):** 623–640.

[26] Gadd GM, Bahri-Esfahani J, Li Q, Rhee YJ, Wei Z, Fomina M and Liang X (2014) Oxalate production by fungi: significance in geomycology, biodeterioration and bioremediation. Fungal Biology Reviews **28(2-3):** 36–55.

[27] Gharieb MM, Sayer JA and Gadd GM (1998) Solubilization of natural gypsum (CaSO$_4$.2H$_2$O) and the formation of calcium oxalate by *Aspergillus niger* and *Serpula himantioides*. Mycological Research **102(7):** 825–830.

[28] Martell AE and Smith RM (2001) Critical Selected Stability Constants of Metal Complexes Database, Version 6.0 for Windows; National Institute of Standards and Technology.

[29] Cruywagen JJ, Rohwer EA and Wessels GFS (1995) Molybdenum (VI) complex formation—8. Equilibria and thermodynamic quantities for the reactions with citrate. Polyhedron **14(23-24):** 3481–3493.

[30] Cruywagen JJ, Heyns JB and van de Water RF (1986) A potentiometric, spectrophotometric, and calorimetric investigation of molybdenum (VI)–oxalate complex formation. Journal of the Chemical Society, Dalton Transactions **9:** 1857–1862.

[31] Haynes WM (2014) CRC Handbook of Chemistry and Physics. 95th Ed., CRC Press, Boca Raton, Florida, USA.

[32] Gopalakrishnan J, Viswanathan B and Srinivasan V (1970) Preparation and thermal decomposition of some oxomolybdenum(VI) oxalates. Journal of Inorganic and Nuclear Chemistry **32(8):** 2565–2568.

[33] Kubicek CP, Röhr M and Rehm HJ (1985) Citric acid fermentation. Critical Reviews in Biotechnology **3(4):** 331–373.

[34] Schatz A (1963) Chelation in nutrition, soil microorganisms and soil chelation. The Pedogenic Action of Lichens and Lichen Acids. Journal of Agricultural and Food Chemistry **11(2):** 112–118.

[35] Iskandar IK and Syers JK (1972) Metal-complex formation by lichen compounds. Journal of Soil Science **23(3):** 255–265.

[36] Pinheiro SMM and Silva MR (2004) Microorganisms and aesthetic biodeterioration of concrete and mortar. pp. 47–54. *In*: Silva MR (ed.). Second International RILEM Workshop on Microbial Impact on Building Materials. RILEM, Paris, France.

[37] Giannantonio DJ, Kurth JC, Kurtis KE and Sobecky PA (2009) Effects of concrete properties and nutrients on fungal colonization and fouling. International Biodeterioration and Biodegradation **63(3):** 252–259.

[38] Fomina M, Podgorsky VS, Olishevska SV, Kadoshnikov VM, Pisanska IR, Hillier S and Gadd GM (2007) Fungal deterioration of barrier concrete used in nuclear waste disposal. Geomicrobiology Journal 24(7-8): 643–653.

[39] Bowen AD, Davidson FA, Keatch R and Gadd GM (2007) Induction of contour sensing in *Aspergillus niger* by stress and its relevance to fungal growth mechanics and hyphal tip structure. Fungal Genetics and Biology 44(6): 484–491.

[40] Maheshwari R (2016) Fungi: Experimental Methods In Biology, Second Edition. Mycology. CRC Press, Boca Rato, FL, USA.

[41] Burford EP, Kierans M and Gadd GM (2003) Geomycology: fungi in mineral substrata. Mycologist 17(3): 98–107.

[42] Money NP (2004) The fungal dining habit: a biomechanical perspective. Mycologist 18(2): 71–76.

[43] Bechinger C, Giebel K-F, Schnell M, Leiderer P, Deising HB and Bastmeyer M (1999) Optical measurements of invasive forces exerted by appressoria of a plant pathogenic fungus. Science 285(5435): 1896–1899.

[44] Gorbushina AA, Krumbein WE, Hamman CH, Panina L, Soukharjevski S and Wollenzien U (1993) Role of black fungi in color change and biodeterioration of antique marbles. Geomicrobiology Journal 11(3): 205–211.

[45] Perfettini JV, Revertegat E and Langomazini N (1991) Evaluation of cement degradation induced by the metabolic products of two fungal strains. Experientia 47(6): 527–533.

[46] Gaylarde CC and Gaylarde PM (2005) A comparative study of the major microbial biomass of biofilms on exteriors of buildings in Europe and Latin America. International Biodeterioration and Biodegradation 55(2): 131–139.

[47] Zhdanova NN, Zakharchenko VA, Vember VV and Nakonechnaya LT (2000) Fungi from Chernobyl: mycobiota of the inner regions of the containment structures of the damaged nuclear reactor. Mycological Research 104(12): 1421–1426.

[48] Giannantonio DJ, Kurth JC, Kurtis KE and Sobecky PA (2009) Molecular characterization of microbial communities fouling painted and unpainted concrete structures. International Biodeterioration and Biodegradation 63(1): 30–30.

[49] McCormack K, Morton LHG, Benson J, Osborne BN and McCabe RW (1996) A preliminary assessment of concrete biodeterioration by microorganisms. pp. 68–70. In: Gaylarde CC, de Sá ELS and Gaylarde PM (eds.). Biodegradation and Biodeterioration in Latin America. Mircen/UNEP/UNESCO/ICRO-FEPAGRO/ UFRGS, Porto Alegre, Brazil.

[50] Lajili H, Devillers P, Grambin-Lapeyre C and Bournazel JP (2008) Alteration of a cement matrix subjected to biolixiviation test. Materials and Structures 41(10): 1633–1645.

[51] Nielsen KF, Holm G, Uttrup LP and Nielsen PA (2004) Mould growth on building materials under low water activities. Influence of humidity and temperature on fungal growth and secondary metabolism. International Biodeterioration and Biodegradation 54(4): 325–336.

[52] Gu J-D, Ford TE, Berke NS and Mitchell R (1998) Biodeterioration of concrete by the fungus *Fusarium*. International Biodeterioration and Biodegradation 41(2): 101–109.

[53] Koval EZ, Serebrenik VA, Roginskaya EL and Ivanov FM (1991) Mycodestructors of building structures of inner premises of food industry enterprises. Microbiologichny Zhurnal 53: 96–103 (in Russian).

[54] Fomina MO, Olishevska SV, Kadoshnikov VM, Zlobenko BP and Pidgorsky VS (2005) Concrete colonization and destruction of mitosporic fungi in model experiment. Mikrobiolohichnyĭ Zhurnal 67(2): 96–104.

[55] Park J-M, Park S-J, Kim W-J and Ghim S-Y (2012) Application of antifungal CFB to increase the durability of cement mortar. Journal of Microbiology and Biotechnology 22(7): 1015–1020.

[56] Park J-M, Park S-J and Ghim SY (2013) Characterization of three antifungal calcite-forming bacteria, *Arthrobacter nicotianae* KNUC2100, *Bacillus thuringiensis* KNUC2103, and *Stenotrophomonas maltophilia* KNUC2106, derived from the Korean islands, Dokdo

and their application on mortar. Journal of Microbiology and Biotechnology **23(9):** 1269–1278.

[57] Park S-J, Park S-H and Ghim SY (2014) The effects of *Paenibacillus polymyxa* E681 on antifungal and crack remediation of cement paste. Current Microbiology **69(4):** 412–416.

[58] Tanaca HK, Dias CMR, Gaylarde CC, John VM and Shirakawa MA (2011) Discoloration and fungal growth on three fiber cement formulations exposed in urban, rural and coastal zones. Building and Environment **46(2):** 324–330.

[59] Shirakawa MA, Aihara EY, Dias CMR, Gaylarde CC and John VM (2008) Fungal colonization on fiber cement exposed to the elements in a tropical climate. *In*: 11th International Conference on Durability of Building Materials and Components, Istanbul, Turkey.

[60] Dadachova E, Bryan RA, Huang X, Moadel T, Schweitzer AD, Aisen P, Nosanchuk JD and Casadevall A (2007) Ionizing radiation changes the electronic properties of melanin and enhances the growth of melanized fungi. PloS ONE **2(5):** 1–13.

[61] Douglas J and Singh J (1995) Investigating dry rot in buildings. Building Research and Information **23(6):** 345–352.

[62] Low GA, Young ME, Martin P and Palfreyman JW (2000) Assessing the relationship between dry rot fungus *Serpula lacrymans* and selected forms of masonry. International Biodeterioration and Biodegradation **46(2):** 141–150.

[63] Palfreyman JW, Phillips EM and Staines HJ (1996) The effect of calcium ion concentration on the growth and decay capacity of *Serpula lacrymans* (Schumacher ex Fr.) Gray and Coniophora puteana (Schumacher ex Fr.) Karst. Holzforschung **50(1):** 3–8.

[64] Connolly JH and Jellison J (1995) Calcium translocation, calcium oxalate accumulation, and hyphal sheath morphology in the white-rot fungus *Resinicium bicolor*. Canadian Journal of Botany **73(6):** 927–936.

[65] Rosato VG and Traversa LP (2004) Lichens on road bridges located in Urban, Rural and coastal environments of Buenos Aires Province, Argentina. pp. 11–18. *In*: Silva MR (ed.). Second International RILEM Workshop on Microbial Impact on Building Materials. RILEM, Paris, France.

[66] Rosato VG (2006) Diversity and distribution of lichens on mortar and concrete in Buenos Aires province, Argentina. Darwiniana **44(1):** 89–97.

[67] Favero-Longoa SE, Castelli D, Fubinic B and Piervittori R (2009) Lichens on asbestos–cement roofs: Bioweathering and biocovering effects. Journal of Hazardous Materials **162(2-3):** 1300–1308.

[68] George RP, Ramya S, Ramachandran D and Mudali UK (2013) Studies on biodegradation of normal concrete surfaces by fungus *Fusarium* sp. Cement and Concrete Research **47:** 8–13.

[69] Rosato VG and Traversa LP (2000) Lichen growth on a concrete dam in a rural environment (Tandil, Buenos Aires province, Argentina). pp. 64–69. *In*: Silva MR (ed.). First International RILEM Workshop on Microbial Impact on Building Materials. RILEM, Paris, France.

[70] Figg J, Bravery A and Harrison W (1986) Covenham Reservoir wave wall—a full-scale experiment on the weathering of concrete. pp. 469–492. *In*: ACI Special Publication 100: The Katherine and Bryant Mather International Conference on Concrete Durability. ACI SP-100-27, American Concrete Institute, Farmington Hills, MI, USA.

[71] Traversa LP, Rosato VG, Pittori CA and Zicarelli S (2001) Biological studies on a concrete dam. Materials and Structures **34(8):** 501–505.

[72] Bertron A and Duchesne J (2013) Attack of cementitious materials by organic acids in agricultural and agrofood effluents. pp. 131–173. *In*: Alexander M, De Belie N and Bertron A (eds.). Performance of Cement-based Materials in Aggressive Aqueous Environments, RILEM State-of-the-Art Report TC 211-PAE. Springer, Dordrecht, Netherlands.

[73] Larreur-Cayol S, Bertron A and Escadeillas G (2011) Degradation of cement-based materials by various organic acids in agro-industrial waste-waters. Cement and Concrete Research **41(8):** 882–892.

[74] Giordani P and Modenesi P (2003) Determinant factors for the formation of the calcium oxalate minerals, weddellite and whewellite, on the surface of foliose lichens. The Lichenologist **35(3)**: 255–270.

[75] Salvadori O and Tretiach M (2002) Thallus-substratum relationships of silicicolous lichens occurring on carbonatic rocks of the Mediterranean region. Bibliotheca Lichenologica **82**: 57–64.

[76] Herdtweck E, Kornprobst T, Sieber R, Straver L and Plank J (2011) Crystal structure, synthesis, aund properties of tri-calcium di-citrate tetra-hydrate $(Ca_3(C_6H_5O_7)_2(H_2O)_2).2H_2O$. Zeitschrift für Anorganische und Allgemeine Chemie **627**: 655–659.

[77] Kaduk J A (2013) Crystal structures of group 2 citrate salts, abstract from ICDD Annual Spring Meetings. Powder Diffraction **28(2)**: 146.

[78] Dyer T (2016) Influence of cement type on resistance to organic acids. Magazine of Concrete Research (in press).

[79] Peynaud E (1984) Knowing and Making Wine. John Wiley, Chichester, UK.

[80] Zuo Y and Hoigné J (1994) Photochemical decomposition of oxalic, glyoxalic and pyruvic acid catalysed by iron in atmospheric waters. Atmospheric Environment **28(7)**: 1231–1239.

[81] Foster JW (1949) Chemical Activities of Fungi. Academic Press, New York, NY.

[82] Guggiari M, Bloque R, Aragno M, Verrecchia E, Job D and Junier P (2011) Experimental calcium-oxalate crystal production and dissolution by selected wood-rot fungi. International Biodeterioration and Biodegradation **65(6)**: 803–809.

[83] Warscheid T and Braams J (2000) Biodeterioration of stone: a review. International Biodeterioration and Biodegradation **46(4)**: 343–368.

[84] Shannon PM, Unnithan V, Bouriak S and Chachkine P (1998) Role for lichen melanins in uranium remediation. Nature **391**: 649–650.

[85] Mottershead D and Lucas G (2000) The role of lichens in inhibiting erosion of a soluble rock. Lichenologist **32(6)**: 601–609.

[86] McIlroy de la Rosa JP, Warke PA and Smith BJ (2012) Lichen-induced biomodification of calcareous surfaces: Bioprotection versus biodeterioration. Progress in Physical Geography **37(3)**: 325–351.

[87] Little B, Staehle R and Davis R (2001) Fungal influenced corrosion of post-tensioned cables. International Biodeterioration and Biodegradation **47(2)**: 71–77.

[88] Strigáč J and Martauz P (2016) Fungistatic properties of granulated blastfurnace slag and related slag-containing cements. Ceramics-Silikáty **60(2)**: 19–26.

[89] Macphee DE and Folli A (2016) Photocatalytic concretes—The interface between photocatalysis and cement chemistry. Cement and Concrete Research **85**: 48–54.

[90] Yu JC-M (2006) Deactivation and regeneration of environmentally exposed Titanium Dioxide (TiO_2) based products. Testing report prepared for the environmental protection department. Chinese University in Hong Kong, Shatin, Hong Kong.

[91] American Concrete Institute (2005) Specifications for Structural Concrete, ACI 301-05. American Concrete Institute, Farmington Hills, MI, USA.

[92] Robinson RF and Austin CR (1951) Effect of copper-bearing concrete on mould. Industrial Engineering Chemistry **43(9)**: 2077–2082.

[93] Dyer TD, Jones MR and Garvin S (2009) Exposure of Portland cement to multiple trace metal loadings. Magazine of Concrete Research **61(1)**: 57–65.

[94] Levowitz LD (1952) Anti-bacterial cement gives longer lasting floors. Food Engineering **24(6)**: 57–60 and 134–135.

[95] Atwan DS (2014) Influence of types anti-fungal admixtures on the concrete mix. Nahrain University, College of Engineering Journal **16(2)**: 180–191.

[96] Do J, Song H, So H and Soh Y (2005) Antifungal effects of cement mortars with two types of organic antifungal agents. Cement and Concrete Research **35(2)**: 371–376.

[97] BASF (2016) MasterLife AMA 100 Integral Antimicrobial Admixture. BASF, Cleveland, OH, USA.

[98] Bolkan SA, Stairiker C, Czechowski MH and Ventura M (2015) Stable aqueous solutions of silane quat ammonium compounds. US 8956665 B2.

[99] Uysal A (2015) Self-cleaning and air cleaning cement. EP 2957546 A1.

[100] Fukushima Y, Shuto Y, Ohinata T and Okouchi S (2008) Antifungal effect of calcined colemanite. Journal of Antibacterial and Antifungal Agents **36(5):** 293–297.

[101] Park S-K, Kim J-HJ, Nam J-W, Phan HD and Kim J-K (2009) Development of anti-fungal mortar and concrete using zeolite and zeocarbon microcapsules. Cement and Concrete Composites **31(7):** 447–453.

[102] Freed WW (1995) Reinforced concrete containing antimicrobial-enhanced fibers. WO/1995/006086.

[103] Ramirez TH and De Leon GD (2004) Concrete-based floors and wall coverings with an antimicrobial effect. US20040121140 A1.

[104] Navas J and Borralleras P (2005) Concrete with antibiotic and antifungal properties. pp. 65–75. *In*: Sánchez Espiniza E and Garcimartin MA (eds.). Proceedings of the Vth International Symposium on Concrete for a Sustainable Agriculture. San Lorenzo de El Escorial, Spain.

[105] Warscheid T (2004) Prevention and remediation against biodeterioration of building materials. pp. 29–36. *In*: Silva MR (ed.). Second International RILEM Workshop on Microbial Impact on Building Materials. RILEM, Paris, France.

[106] Eyssautier-Chuine S, Gommeaux M, Moreau C, Thomachot-Schneider C, Fronteau G, Pleck J and Kartheuser B (2014) Assessment of new protective treatments for porous limestone combining water-repellency and anti-colonization properties. Quarterly Journal of Engineering Geology and Hydrogeology **47(2):** 177–187.

[107] Highways Agency (2007) Inspection Manual for Highway Structures: Vol. 1: Reference Manual. Highways Agency, The Stationery Office, London, UK.

[108] May E, Lewis FJ, Pereira S, Tayler S, Seaward MRD and Allsopp D (1993) Microbial deterioration of building stone—a review. Biodeterioration Abstracts **7(2):** 109–123.

[109] Dyer TD (2014) Concrete Durability. CRC Press, Boca Raton, FL, USA.

[110] Historic Scotland, Biological Growth on Masonry: Identification and Understanding. Historic Scotland, Edinburgh, UK.

[111] Brown SK (1982) Fibre release from corrugated asbestos cement cladding due to weathering and cleaning. CSIRO Division of Building Research, Highett, Australia, 35 p.

[112] Martin AK, Johnson GC, McCarthy DF and Filson RB (1992) Attempts to control lichens on asbestos cement roofing. International Biodeterioration and Biodegradation **30(4):** 261–271.

Chapter 5

Plants and Biodeterioration

5.1 Introduction

The kingdom of plants incorporates a significant variety of organisms in terms of both size and form. Their two common features are the use of photosynthesis as a means of obtaining energy and the material that makes up their cell walls: cellulose.

The interaction of plants with concrete structures in manners which can be problematic are both chemical and physical. However, most physical damage primarily derives from larger plants. For this reason, these divisions of the plant kingdom are discussed below. Whilst usually not harmful to concrete, the development of moss and similar plants can sometimes play a role in the establishment of other plants on concrete surfaces, and the division of plants in which mosses are placed (*Bryophyta*) is also covered.

5.1.1 Algae

The term algae covers a wide range of organisms, whose similarity and distinction from other organisms is not easy to define. One definition is that they obtain energy by photosynthesis using chlorophyll, and that they have reproductive cells that are not protected by layers of sterile cells [1]. Algae can be both unicellular and multicellular, with some of the multicellular species, such as seaweeds, being able to grow to a considerable size. Both unicellular and multicellular algal forms are most probably present in the photograph in Figure 5.1. Many of the unicellular species are motile.

When algae are discussed in this chapter, it will only be in the context of eukaryotic algae—organisms whose cells contain a nucleus and other organelles contained within membranes. Cyanobacteria are prokaryotic organisms—lacking a nucleus bound within a membrane—which are often also included within the grouping of algae. Cyanobacteria are covered in Chapter 3.

Figure 5.1 Algae in the form of an algal biofilm and seaweed at the waterline of a concrete bridge.

Plants and most algae undergo oxygenic photosynthesis—they are photoautotrophs utilising sunlight as a source of energy, carbon dioxide as a source of carbon, and water as the electron donor:

$$photons + 2H_2O + CO_2 \rightarrow [CH_2O] + O_2 + H_2O$$

where $[CH_2O]$ is a carbohydrate—not necessarily with this specific chemical formula—such as glucose. In fact, algae can be both autotrophic or heterotrophic, with a number being capable of obtaining energy through both mechanisms (mixotrophic).

The carbohydrate acts as a means of storing energy, which can be accessed by processes of cellular respiration where carbohydrate is oxidised. However carbohydrates may also be used as the building blocks for cell walls or precursor molecules in the biosynthesis of other molecules needed by the organism. In the case of algae, cell membranes are made from fibrillar components embedded in an amorphous matrix [1]. The fibrillar components are most commonly made from cellulose (polymerised glucose), but can sometimes be polymers of mannose (mannan) or xylose (xylan). The amorphous matrix is also composed of polysaccharides in the form of alginic acid, fucoidin, agar and carrageenan.

Algae can be categorised in terms of the substances used for photosynthesis. Green algae use Chlorophyll *a* and *b*, whilst red algae use chlorophyll *a* and *c* plus phycobilins. Brown algae use chlorophyll *a* and *c*, and fucoxanthin.

Because of the presence of chlorophyll, the presence of algal biofilms is usually fairly evident, as a result of their colour. However, it should be noted that cyanobacteria (see Chapter 3) also contain similar pigments, and are often encountered alongside algae.

The larger, multicellular algae more closely resemble higher plants, in that they possess comparable structures. The structural features that make up multicellular algae such as seaweeds are shown in Figure 5.2. These features include laminae, which are flat structures approximately resembling leaves. Seaweeds may also possess air bladders—which are used to keep the plant afloat—and a stem known as a stipe. It should be stressed that significant variation exists between different genera and species with regards to the shape, location and prominence of these features. For instance, many seaweeds do not possess a stipe, whilst air bladders can also be located as features of the laminae.

Seaweeds do not, however, possess roots: they are anchored to the substrate on which they grow by a structure known as a 'holdfast' which may take many forms depending on the nature of the substrate. Where algae are growing in granular media it may take the form of a bulb, whereas in substrates made of finer particles, it may more closely resemble a rootlike structure. On solid surfaces, the holdfast takes the form of a flat structure which is attached to the surface using adhesive substances secreted by the organism.

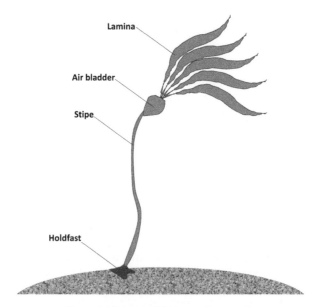

Lamina

Air bladder

Stipe

Holdfast

Figure 5.2 Structural features of seaweed.

5.1.2 Vascular plants

Vascular plants are multicellular organisms which possess two tissues which allow the distribution of substances to all of their cells. The first of these is the xylem which transports water and nutrients from the roots to the rest of the plant. The second is the phloem which is used to transport the products of photosynthesis from photosynthesising cells to other cells, including those in the roots. The majority of photosynthesis is conducted in cells within the leaves which normally possess a large surface area which is able to maximise the number of cells illuminated by sunlight.

An extremely generic schematic diagram of some of the key features of a vascular plant is provided in Figure 5.3. However, the description above avoids one extremely important aspect of the vascular plants, which is the sheer diversity of scale and form that exists within the vascular plants.

The cell walls of plants is composed largely of cellulose. The xylem also contains lignin, which imparts rigidity to varying degrees.

5.1.3 Bryophyta

Bryophytes are a division of plants which differ from the higher plants in that they lack a vascular system—they lack a xylem and phloem for

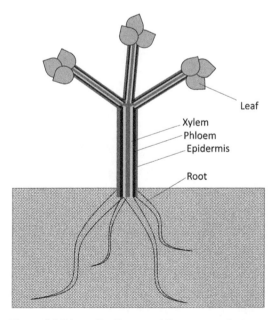

Leaf
Xylem
Phloem
Epidermis
Root

Figure 5.3 Schematic diagram of the common features
of a vascular plant.

the distribution of water. Of most significance—within the context of this chapter—are the mosses (Bryophyta)—Figure 5.4.

Mosses are found in both terrestrial and aquatic environments. They are multicellular and consist of a short stem with simple leaves, and are anchored to their substrate by thread-like rhizoids. Rhizoids appear to be purely employed for this purpose and, unlike the roots of vascular plants, are unable to absorb water and nutrients. However, they can, in some cases, penetrate porosity in stone or concrete.

5.2 Plant Growth and Reproduction

The growth of vascular plants occurs through a process of cell division, taking the form of the development and growth of shoots and leaves, the radial growth and thickening of roots and the stem/trunk.

Figure 5.4 Moss on a concrete wall.

Bryophyte growth takes the form of progressive, lengthening of the stem, the growth of new leaves, further development of rhizoids and the development of sporophytes on female plants (discussed below).

Growth in the case of algae is very much dependent on the type and size of the organism. Whilst unicellular algae effectively 'grow' by reproduction, such organisms will often form highly organised structures with algae of the same species. For instance, non-motile unicellular algae may form filamentous structures-long strings of cells–or palmelloid structures– groupings of cells embedded in a mucilage. Similarly, motile algae may form small, mobile colonies. The multicellular algae grow in a manner more like that of vascular plants, although the absence of roots means that growth is limited to the other parts of the organism.

Reproduction in plants can occur via a number of processes, with a diversity of variations in how these processes are achieved. Given the focus of this book, it is of limited value to discuss these processes in too much detail. However, the reproductive processes of algae, vascular plants and bryophytes are outlined below, since there are aspects which will help in understanding how concrete surfaces can become colonised by plants.

5.2.1 Algal reproduction

The means by which algae reproduce are diverse and complex. However, they can be grouped into three general forms: vegetative, asexual and sexual. Vegetative reproduction involves either cell division or division of multicellular algae through various mechanisms. Asexual reproduction employs spores, or similar entities, which grow into a plant genetically identical to the parent plant. Sexual reproduction involves either isogamy (where algal cells from different individuals form similar motile gametes which fuse together) or heterogamy (where gametes with very different size and/or characteristics are produced).

5.2.2 Higher plant reproduction

Plants can reproduce sexually and asexually. Sexual reproduction involves the transfer of male gametes in the form of pollen (in the case of flowering plants) to female ovules, also located within the flowers. This transfer is often achieved through another organism that acts as a pollinator. The pollinator is most commonly an insect, but other animals can also be employed. Fertilization leads to the formation of seeds within fruit. The manner in which seeds are dispersed is of some relevance to the topic under discussion, with wind and animal dispersion being two common means by which this is achieved. Seeds will germinate when they experience appropriate temperatures and levels of moisture and oxygen.

In the case of the ferns, reproduction occurs through a slightly different mechanism. The fern (in its 'sporophyte' form) develops leaves known as sporangium which produce spores. These spores are released and dispersed by processes such as the action of wind, animals, gravity and the movement of water. Once a spore encounters suitable conditions, it germinates to form a small plant—a 'gametophyte'—which possesses two organs: the antheridia and the archegonia. The antheridia produces motile sperm, whilst the archegonia contains female egg cells. The sperm can swim away from the antheridia if water is present, for instance, as a result of rainfall. Once a sperm fertilizes an egg cell, a zygote forms which grows into a sporophyte (essentially what would normally be considered the 'fern' stage of the lifecycle).

5.2.2 Bryophyte reproduction

Mosses reproduce via spores. When a spore germinates it develops into either a male or female gametophyte. The male form develops antheridia, whilst the female plant develops archegonia, not unlike those of the ferns. In suitable conditions, motile sperm from the antheridia on a male plant to the egg-containing archegonia on a female plant, leading to fertilization. This leads to the growth of a sporophyte, which is an entity completely separate to the gametophyte on which it forms, but partly dependent on the gametophyte for nutrients. The sporophyte develops from within the antheridia, growing upwards from the top of the female gametophyte. A capsule is located at the end of the sporophyte which eventually ripens and releases spores which are usually dispersed by the wind. Once suitable conditions are encountered by the spore, it germinates, firstly developing as a protonema, which will ultimately grow to form a gametophyte.

In all of the above descriptions of the means by which plants reproduce, the fundamental theme that is of significance with regards to the appearance and growth of plants on concrete surfaces is that seed and spore formation plays a frequent role, and that many of these are dispersed by the wind or animals in such a way that the appearance of plants in relatively inaccessible locations on buildings is perfectly possible. This aspect of the life-cycle of plants will be explored further in the next section, where concrete as an environment for plant life is examined.

5.3 Concrete as an Environment for Plant Life

The growth of plants on concrete is a process which normal requires a number of conditions to converge. These include the availability of water, the chemical nature of the concrete surface, and the availability of nutrients. All of these aspects are discussed in this section. However, before they can be discussed, it is important to consider the processes which allow plants to reach—sometimes very remote—locations on a concrete structure.

5.3.1 Steps towards plant colonisation of concrete surfaces

The manner in which concrete surfaces are colonised by plants varies depending on nature of the organism. It is again useful, therefore, to divide the discussion in terms of whether algae, vascular plants or bryophytes are involved.

Algae

For algae to be brought into contact with a concrete surface, there must be some movement of algae-bearing water past the surface. The rate of flow plays an important role in the rate at which algae become attached to this surface. A study which examined rates of attachment of a number of species of algae delivered to different surfaces by flowing water travelling at different rates found that slower rates favoured colonisation [2].

The same study also found that attachment rates were dependent on the nature of the surface. However, the findings are somewhat counterintuitive, with algae often attaching with greater efficiency to hydrophobic surfaces with low free surface energies.

It has been proposed that the attachment of algae is the result of extracellular polymeric substances (EPS) produced by these organisms. This is supported by research which examined the nature of the polysaccharides produced by a number of species of algae isolated from building façades [3]. The polymers were categorised in terms of whether they were released into the environment surrounding the cell (released polysaccharides, RPS), or were retained as the outer envelope of the cells (capsular polysaccharides, CPS). The polymers were found to be anionic in nature and, in some cases, reduced the surface tension of water. The kinetics of CPS adsorption onto a glass surface and silicon wafers coated with a hydrophobic agent were characterised. It was found that, whilst the rates of adsorption were similar initially, the hydrophobic surface ultimately adsorbed more of the polysaccharides.

In the case of the larger multicellular algae, colonisation is likely to be initiated by a spore becoming attached to the surface.

Vascular plants

It has been proposed that higher plants may become established on concrete surfaces through two mechanisms [4].

The first mechanism occurs on both horizontal and vertical surfaces and involves seeds of plants becoming lodged in cracks or holes in the surface.

The second mechanism requires the establishment of moss (or potentially other *Bryophyta* members) on a horizontal concrete surface (see Figure 5.5). The moss has the effect of trapping dust and other particulates carried by the wind leading to a situation where a sufficient quantity of substrate is formed to support the establishment of higher plants with roots.

Only certain plants are suited to the colonisation of a concrete surface. In their review of the damage that plants are capable of doing to historic

Figure 5.5 Moss on a horizontal surface providing a substrate for vascular plant growth.

stone buildings, Lisci *et al.* [4] draw up a list of characteristics likely to favour a given plant in colonising such surfaces:

Suitable seeds: only smaller seeds are likely to be able to become lodged in small cracks and holes. However, if the surface in question is vertical or above ground level, there must also be a means of the seed reaching such locations. Thus, plants whose seeds are adapted to be carried by the wind, or are present in fruits which are eaten by birds are clearly more likely to be found on such surfaces.

Drought resistance: seeds which are able to germinate in the presence of only small quantities of water, and plants which are able to grow under similar conditions are likely to be less affected by the fluctuations in moisture which are likely to be encountered on a stone or concrete surface.

Appropriate means of reproduction: plants which reach sexual maturity rapidly are more likely to grow the population of plants occupying a surface. Additionally, if plants possess a means of vegetative production, this will also assist in this.

Suitable rooting mechanisms: plants which are deep rooting and whose roots are particularly suited to growth through rock and similar materials are likely to have an advantage on stone and concrete surfaces.

The same researchers devised a list of European plants which were known to populate historic stone buildings, along with their propensity to cause damage. Table 5.1 contains plants from this list which were also identified as being harmful to stone structures, plus some other species from other sources. The criteria for selecting harmful species was formation of woody stems and roots likely to cause damage to building as they grew in a radial direction. Added to this table are seed sizes and dispersal characteristics, whether the plant is calcicole or calcifuge, and its preferred pH conditions. Table 5.2 lists plants likely to be damaging from a list of plants found growing on historic structures in tropical regions compiled by Kumar and Kumar [7].

It should be stressed that, as a result of the nature of the main sources of the identified species in Tables 5.1 and 5.2, it is likely that the list is somewhat biased towards Mediterranean and Indian species. However, the list is nonetheless useful, in that it allows the consideration of what types of plants are most likely to grow on a concrete structure.

The first consideration is what types of seed are most likely to become lodged in a concrete surface. Given that concrete surfaces are typically relatively smooth, it is likely that smaller seeds will have a greater chance of locating themselves in a cavity such as a blowhole or crack. It is unlikely that many substantial cracks will exist shortly after construction, and so other forms of deterioration may need to occur prior to the establishment of vascular plants by this route. The thousand seed weight provides some indication of the relative size of the seeds in the Tables. The density of seeds from different genera and species can differ to quite a large extent, and so attempting to translate thousand seed weight into linear dimensions is not advisable. However, it is safe to conclude that the seeds with the lower thousand seed weights are more likely to locate themselves in the types of crack or recess that might be present on a concrete surface.

Additionally, whilst it is evident from the table that some plants can grow in relatively alkaline conditions, the extremely high pH of relatively young concrete is likely to be excessive. This is discussed further in Section 5.3.2. It is also worth noting that the vast majority of the European plants in Table 5.1 are calcicole in nature, meaning that they favour substrates in which calcium is present (see Section 5.3.3).

Table 5.1 Vascular plants likely to be harmful inhabitants of European historic structures and walls [4, 5, 6].

Order	Family	Genus	Species	Common Name	Dispersal	Thousand Seed Weight, g	Calcicole/ Calcifuge	Optimum pH	References
Apiales	Araliaceae	*Hedera*	*helix*	Ivy	animal	69.30	Calcicole	6.0–7.5	[8, 12]
Asterales	Compositae	*Dittrichia*	*viscosa*	Woody Fleabane		0.38	Adaptable ecotype	–	[8, 17]
Brassicales	Capparidaceae	*Capparis*	*spinosa*	Caper	animal	9.00	Calcicole	7.5–8.0	[8, 10, 16]
	Brassicaceae	*Erysimum*	*cheiri*	Wallflower	wind	1.40	Calcicole	5.5–7.5, but can tolerate higher pH	[8, 11, 12]
		Matthiola	*incana*	Stock	wind	1.62	Calcicole	6.0–7.5	[8, 12]
Caryophyllales	Polygonaceae	*Fallopia*	*japonica*	Japanese Knotweed	wind/animal	2.065	–	3.0–8.5	[8, 15, 18]
Celastrales	Celastraceae	*Euonymus*	*europaeus*	Spindle-tree	animal	40.60	Calcicole	6.0–8.0	[8]
Dipsacales	Caprifoliaceae	*Sambucus*	*nigra*	Elderberry	animal	26.00	Calcicole	5.5–6.5	[8, 12]
	Valerianaceae	*Centranthus*	*ruber*	Red Valerian	wind	1.77	Calcicole	5.0–8.0	[8]
Fabales	Leguminosae	*Robinia*	*pseudoacacia*	False acacia/ Black locust	Gravity/root suckers	18.00–20.00	Calcicole	4.6–8.2	[8, 13, 14]

Order	Family	Genus	species	Common name	Dispersal		Soil preference	pH	References
Lamiales	Oleaceae	*Syringa*	*vulgaris*	Lilac		5.60	Calcicole	6.0–7.5	[8, 12]
		Fraxinus	*excelsior*	Ash	wind/gravity	62.6	Calcicole	–	[8]
	Buddlejaceae	*Buddleja*	–	Buddleja/Buddleia		0.01–1.62	Calcicole	4.0–6.0	[12]
Pinales	Taxaceae	*Taxus*	*baccata*	Yew	animal	60.5	Calcicole	5.5–7.5	
Polypodiales	Dryopteridaceae	*Dryopteris*	*filix-mas*	Male fern		–	–	–	
Ranunculales	*Ranunculaceae*	*Clematis*	*vitalba*	Old man's beard	wind	2.10	Calcicole	5.5–8.0	[8, 15]
Rosales	*Ulmaceae*	*Celtis*	*australis*	Nettle-tree	animal	203.70	Calcicole	6.0–7.8	[8, 13]
		Ulmus	*minor*	Elm	wind	5.14	Calcicole	5.5–8.5	[8, 13]
	Moraceae	*Ficus*	*carica*	Fig	animal	1.30	No preference	5.5–8.5	[8, 9, 13]
	Rosaceae	*Rubus*	–	Bramble	animal	0.028–10.40	Calcicole	5.0–6.0	[8, 12]
	Rhamnaceae	*Rhamnus*	*alaternus*	Buckthorn	animal	20.27	Calcicole	–	[8]
	Uricaceae	*Parietaria*	*judaica*	Pellitory of the wall	wind/water/animal	0.247	Calcicole	–	[8]
Sapindales	Simaroubaceae	*Ailanthus*	*altissima*	Tree of Heaven	wind	29.40	Calcicole	5.5–8.5	[8, 13]
	Sapindaceae	*Acer*	*pseudoplatanus*	Sycamore	wind	97.00	–	–	[8]

Table 5.2 Vascular plants identified as being harmful inhabitants of historic structures in tropical regions [7]. Thousand seed weight data from [8].

Order	Family	Genus	Species	Common Name	Dispersal	Thousand Seed Weight, g
Brassicales	Capparaceae	Capparis	flavicans	–	–	[genus range = 3.54–220.00]
			zeylanica	Ceylon caper	–	
Fabales	Fabaceae	Vachellia	nilotica	Gum Arabic tree	–	119.00
		Albizia	lebbeck	Lebbek tree	–	123.00
		Senna	occidentalis	Senna coffee	–	17.00
		Dalbergia	sissoo	North Indian rosewood	–	41.80
Gentianales	Apocynaceae	Calotropis	procera	Apple of Sodom	wind	9.57
Lamiales	Lamiaceae	Leucas	biflora	Two-flowered leucas	–	[genus range = 0.27–2.60]
Lamiales	Orobanchaceae	Lindenbergia	indica	Nettle-Leaved Lindenbergia	–	0.041
Malpighiales	Euphorbiaceae	Croton	bonplandianus	Jungle tulsi	–	7.77
Myrtales	Lythraceae	Woodfordia	fruticosa	Thathirippoovu	–	0.28
Rosales	Moraceae	Ficus	benghalensis	Indian banyan	–	1190.00
			religiosa	Bodhi tree	–	0.54
			rumphii	–	–	–
	Ulmaceae	Holoptelea	integrifolia	Indian elm	wind	34.00
	Rhamnaceae	Zizyphus	jujuba	Jujube	–	
Sapindales	Meliaceae	Azadirachta	indica	Neem	–	241.00

5.3.2 pH

As for the other microorganisms discussed in previous chapters, algae are generally sensitive to pH, with acidic and alkaline environments limiting growth. This is shown in Figure 5.6 which shows the growth rate of populations of the freshwater algae *Cryptomonas*.

The sensitivity to pH varies between genera and species, and there are a number of acidophilic algae. Alkaliphilic species, whilst seemingly less common, are also encountered [20], but the species typically encountered growing on concrete do not appear to have this characteristic.

Normally, maximum rates of algal growth tend to be observed around a pH of 8, meaning that relatively young concrete does not represent a friendly environment. However, carbonation and other processes which reduce the pH of the cement matrix mean that this situation is usually a temporary one. Figure 5.7 shows the progressive growth of populations of *Klebsormidium flaccidum* on mortar surfaces in both carbonated and uncarbonated conditions [21]. Complete occupancy of both types of mortar eventually occurs, but this occurs much faster for the carbonated surface. The experiment involved periodic trickling of a suspension of algae over the mortar surfaces alongside periodic illumination with lamps to provide a means of photosynthesis. Thus, the eventual occupancy of the uncarbonated surface was probably the result of both leaching of alkaline species from the mortar and carbonation as a result of exposure to the air between these periods of exposure to algae.

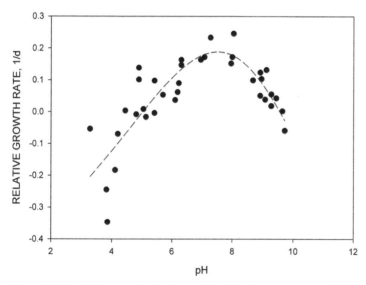

Figure 5.6 Population growth rate of a species of *Cryptomonas* versus pH [19].

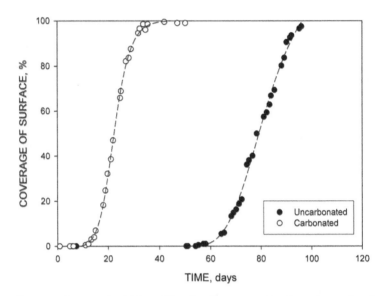

Figure 5.7 Coverage by *Klebsormidium flaccidum* versus time for carbonated and uncarbonated mortar surfaces [21].

Mosses typically prefer more acidic conditions than most plants. Figure 5.8 shows the length of growth of the protonema (the early initial structures formed by moss after germination—see Section 5.2) of the moss *Calymperes erosum*. An optimum growth rate occurs around a pH of 4.5. However, there are a number of alkaliphilic species which can tolerate a somewhat higher pH. For instance, *Scorpidium cossonii* has been shown to favour environments with a pH of 8.30 and tolerating conditions greater than pH 9 [22]. Other research has reported moss growing on stone debris from walls with a pH of around 10 [6].

The preferred pH conditions of vascular plants is highly species-dependent. However, examining the preferred pH conditions of plants identified as being harmful to buildings in Table 5.1, it is evident that the midpoint of the range is typically located close to pH neutral conditions. Some caution should be employed in interpreting these preferred pH ranges, since plants are normally able to grow in more alkaline conditions to the upper limit, albeit at a limited rate. Thus, if the shape of the plot of shoot and root mass development by buddleia plants versus pH (Figure 5.9) is examined, it is evident that growth is still likely beyond a pH of 8.0. Nonetheless, as for the other examples discussed above, mature—and, hence, carbonated concrete—is likely to better support growth.

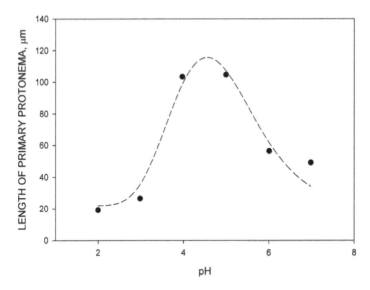

Figure 5.8 Length of primary protonema grown from germinated *Calymperes erosum* spores after 11 days under different pH conditions [23].

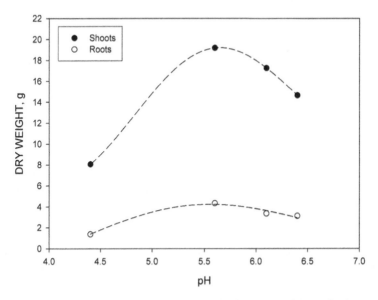

Figure 5.9 Dry mass of the shoots and roots versus pH for *Buddleia davidii* plants 60 days after being transplanted to a pine bark medium. pH differences were achieved through amendments of dolomitic lime [24].

5.3.3 Nutrients

The essential macronutrient elements for plants are carbon, hydrogen, oxygen, phosphorus, nitrogen, sulphur, potassium, calcium and magnesium [25]. Hydrogen is obtained from water, which will be discussed in a later section.

In the higher plants and many algae, carbon is obtained from carbon dioxide, in which case, the shortage of sources of carbon that many organisms living on concrete encounter will not be an issue. Carbon dioxide is taken in by plants through the leaves, and through the cell membranes by algae.

All of the remaining nutrients must be taken up by the roots of vascular plants.

Being located underground, roots cannot undergo photosynthesis. Thus, the energy for root growth is obtained from the sugars produced by the leaves via cellular respiration. This requires oxygen, resulting in the production of carbon dioxide. For plants growing on a concrete surface, the availability of oxygen to roots is not an issue, unless the surface becomes submerged by water. However, where roots of trees are growing beneath concrete pavements, the lack of oxygen can have the effect of directing the growth of roots upwards, which can present problems. This issue is discussed later in this chapter.

Nitrogen, on the other hand, is likely to be in shorter supply, since this usually needs to be in the form of nitrate or ammonium ions for higher plants. There is very little nitrogen in any form in concrete, and so it must come from an external source. However, plants may also obtain the element from nitrogen gas, through the process of nitrogen fixation: the conversion of nitrogen gas to ammonia.

Nitrogen fixation employs microorganisms as a means of converting atmospheric nitrogen. These organisms may either be free and living in close proximity to the plant roots, or for some plants, symbionts held within root nodules or sometimes other structures within the plant. The micro-organisms involved are most commonly bacteria, but can also be cyanobacteria. Ammonia may subsequently be oxidised to nitrite, and then nitrate, by nitrifying bacteria (See Chapter 3).

Algae must also obtain nitrogen in the form of nitrates of ammonia. However, it is notable that populations of algae often develop co-existing with those of cyanobacteria [26] whose ability to fix nitrogen is presumably beneficial.

The availability of sulphur, potassium and magnesium have been dealt with in some detail in Chapter 3. However, it is worth revisiting the availability of calcium in concrete due to its abundance, and also because of its importance as a nutrient to a number of plants.

The higher plants can broadly be categorised as either *calcicoles* or *calcifuges*. Calcifuges grow in soils with a low calcium concentration and, in some cases, may grow less well if this concentration is increased. Calcicoles, on the other hand, grow in soils with high calcium concentrations and display symptoms of calcium deficiency when this element is scarce [27]. Table 5.1 also identifies whether each plant listed is calcicole or calcifuge, with the vast majority falling into the former category. Thus, it can be expected that most of the plant species growing on concrete structures will be calcicole in nature.

One of the elements present in concrete which has the potential to make concrete a less hospitable environment to plants is aluminium. Aluminium-whilst a potentially beneficial nutrient in small quantities—is toxic to plants [28]. Concentrations sufficient to cause detrimental effects are not normally encountered under pH conditions greater than 5, making this a lesser threat to plants growing on concrete. However, it is worth noting that one defence that plants seemingly employ to combat aluminium toxicity is the exudation of organic acids [29]. These acids include citric, malic, oxalic, tartaric and fumaric acid [30], but others may also be exuded. It is proposed that these acids form complexes with aluminium in solution, rendering it less available for uptake by the plant and, hence, less toxic.

However, whilst increased aluminium concentrations (and sometimes concentrations of other metals) stimulate the release of these acids, this is also stimulated by the *absence* of phosphorus. This indicates that the acids most probably have two roles: limiting aluminium toxicity and increasing phosphate availability [30]. It has been proposed that the release of these acids solubilises phosphate through the formation of stronger complexes than phosphate with aluminium, iron and calcium and adsorbing on charged surfaces that would otherwise attract phosphate.

Thus, whilst exudation of organic acids as a result of high aluminium concentrations is an unlikely requirement in most cases for plants growing on concrete, the typically low concentrations of phosphate present in cement and aggregate might be sufficient to stimulate this. Given the damaging effect that citric acid can have on concrete (see Chapter 4), release of acid from roots has the potential to contribute towards deterioration.

Calcicole plants typically produce larger quantities of organic acid. A series of experiments which examined both the nature and concentrations of different organic acids produced from the roots of two calcicole and two calcifuge species found that acetic, lactic, malic, succinic, oxalic and citric acid were produced by the two calcicole species (*Gypsophilia fastigata* and *Sanguisorba minor*) [31]. Whilst the calcifuge species (*Deschampsia flexuosia* and *Viscaria vulgaris*) did exude most of these acids to some extent, the quantities were much less. *Gypsophilia fastigata* also produced significant quantities of tartaric acid. Further experiments revealed that oxalic acid had the effect of solubilising phosphate, whilst citric acid solubilised iron and manganese from the soils used.

Heterotrophic algae have also been found to produce organic acids. These include formic, pyruvic, lactic and succinic acids. One study examining acid formation by three types of green algae (*Chlorella vulgaris, Scenedesmus basilensis* and a species of *Chlamydomonas*) [32]. Under aerobic conditions with glucose as the source of energy, pyruvic acid was produced in the largest quantities by *Chlorella* and *Scenedesmus*. *Chlamydomonas* produced formic acid in larger concentrations under aerobic conditions, albeit in much smaller concentrations overall. Under anaerobic conditions, all three algal cultures produced lactic acid. These acids are most likely formed as products of metabolising glucose, rather than as a means of adjusting the surrounding conditions, as in the case of plant roots. However, it should be noted that even under autotrophic conditions (i.e., in the absence of glucose) the acids were all produced by the mixotrophic *Chlorella vulgaris*, although in much smaller concentrations.

One possible example of exudation of organic acids by algae being used functionally is in the case of cryptoendolithic algae which appear to live in layers just beneath rock surfaces as a means of limiting the harmful effects of extreme environmental conditions such as high or low ambient temperatures and low humidity [33]. It has been proposed that these recesses—and possible subsequent biodegradation of the stone—have been formed through the exudation of acids [34]. This strategy demands that the rock is sufficiently optically clear to permit sunlight to pass through the surface to allow the algae to photosynthesize, with sandstones proving ideal [33]. Whether typical concrete formulations contain suitable aggregates to permit this is uncertain.

One nutrient required by a specific group of algae is silicon. The group in question is the diatoms, which are unicellular algae found in both freshwater and marine environments. The silicon is used by diatoms to form a porous, symmetrical cell wall made of hydrated silica known as a frustule. There is obviously a quantity of silicon present in cement and aggregate. Generally the Si in the cement matrix will be present as CSH gel, and will be more soluble than that present in the aggregate. However, the solubility of Si in CSH is still low. Whilst, as will be seen in Section 5.4, diatoms may be found in close proximity to concrete structures, diatoms do not attach to surfaces, but instead remain suspended in water. Moreover, there is no evidence that diatoms produce substances to render Si more available. Thus, concrete does not provide a habitat for diatoms, and is unlikely to be affected by their proximity.

It is generally believed that the rhizoids of moss solely perform the function of anchors. Rhizoids appear to be capable of penetrating rock, mortar and concrete, but seemingly only if the porosity is sufficient to allow this [35]. Many bryophyta—especially of the *Sphagnum* genus—produce quantities of secondary metabolites including phenolic compounds and carbohydrates [36]. Only some of these substances are released by the moss and many of the compounds are present as soluble constituents in the cells or as components of the cell walls.

The phenolic compounds appear to play a number of roles. Firstly, they protect the moss from being eaten by herbivores, through an unpleasant taste, toxicity and an ability to 'mask' cellulose for digestion. Additionally, it is believed that the compounds can inhibit the growth of vascular plants. Finally their presence tends to acidify the soil, to produce conditions which are optimal for the moss.

The carbohydrates, which include uronic acids, appear to be able to undergo cation exchange of protons with nutrient ions including calcium and magnesium. Additionally, cation exchange leads to the release of protons which further acidify the substrate.

It would appear that a lower calcium concentration in the substrate promotes the development of a higher cation exchange capacity in the cell walls [37]. Sphagnum mosses are also known as peat mosses, and are encountered in peat bogs. However, other moss species also form similar compounds, albeit often in lesser quantities.

5.3.4 Concrete as a habitat for plants

Algae

The availability of moisture is a fundamental requirement of algae, and as a result it might be expected that water/cement ratio is likely to play an important role in defining how hospitable a concrete surface is. This has already been seen in Chapter 3 (Figure 3.3) for mixed communities of algae and cyanobacteria. The only study in which algae were the sole occupier, the experimental programme utilised mortars with relatively high ratios and used test conditions which were unlikely to allow the mortar surface to dry to any great extent [21]. As a result, rates of growth were almost identical. However, it is reasonable to assume that a higher water/cement ratio will produce a surface which is more amenable to algal growth.

The same study investigated the influence of surface roughness and found a rougher surface allowed faster rates of colonisation by algae.

Vascular plants

Water is also of key importance to vascular plants, and so the presence of moisture within the concrete surrounding a plant is conducive to survival and growth. Whilst it is likely that concrete with a high water/cement ratio would provide a larger reservoir of water to plants—as for other organisms—this effect has not been investigated. However, it is likely that other factors will also play a role in providing a moist environment. In particular, the establishment of bryophytes (see 5.3.1) is likely to provide the means for the retention of water close to the surface.

5.4 Plants on Concrete

Whilst a significant amount of research has been conducted into the identification of algae occupying concrete surfaces, data on bryophytes is limited to a small number of studies, whilst coverage of vascular plants is non-existent. In the case of vascular plants, there is at least the data provided in Table 5.1, for other substrates comparable in some regards to concrete. This section focusses, therefore, on algae and bryophytes.

5.4.1 Algae

Table 5.3 lists the species identified by studies examining algal growth on concrete and mortar surfaces either in the field or under laboratory conditions.

5.4.2 Bryophytes

There have been a limited number of studies on the characterisation of bryophyte species growing on concrete surfaces. This limitation is further compounded by the fact that these studies are largely limited to a relatively narrow geographical area. Nonetheless, many of these studies have studied extensive sites and have been extremely thorough in the characterisation process. Thus, whilst the list of species provided in Table 5.4 is a long one, it must be recognised that the full list of species that have the potential to grow on concrete is considerably larger.

5.5 Damage from Root Growth

The main process which leads to damage to structures by vascular plants is the radial growth of roots and woody stems in confined spaces within which a plant is growing. This type of physical damage is, of course, not unique to concrete structures, and reports of plant damage to concrete structures is not particularly well reported in academic literature, although its occurrence is without question. For this reason, aspects of the discussion below have been extended to include sources involving construction materials other than concrete, where the findings are still relevant.

The pressures exerted by roots as they grow can act both radially (outwards) and longitudinally (at the root tip). The magnitude of these

Table 5.3 Algae identified growing on concrete surfaces, or successfully grown in the laboratory.

Genus	Species	Class/Order/Family	Habitat	Location	References
Aglaothamnion	–	**Florideophyceae/** Ceramiales/ Callithamniaceae	Marine	Finland	[46]
Apatococcus	*lobatus*	**Chlorophyta/** Trebouxiophyceae/ Chlorellales	Freshwater	France	[26]
Ascophylum	*nodosum*	**Phaeophyceae/** Fucales/Fucaceae	Marine	Fylde coast, UK	[39]
Bracteacoccus	–	**Chlorophyta/** Chlorophyceae/ Chlorococcales	Freshwater	France	[26]
Ceratium	–	**Dinophyceae/** Gonyaulacales/ Ceratiaceae	Marine		[46]
Chlorella	*ellipsoidea*	**Chlorophyta/** Trebouxiophyceae/ Chlorellales	Freshwater	France/ Laboratory	[26, 43]
	homosphaera			France	[26]
	minutissima			Boersch, France/ France	[3, 26]
	vulgaris			Laboratory	[41]
	mirabilis			Laboratory/ France	[42, 26]
Chlorosarcinopsis	*eremi*	**Chlorophyta/** Chlorophyceae/ Chlorosarcinales	Freshwater	France	[26]
	minor			France	[26]
Choricystis	*minor*	**Chlorophyta/** Chlorophyceae/ Chlamydomonadales	Freshwater	France	[26]
	chodatii			France	[26]
Coccobotrys	*verrucariae*	**Chlorophyta/** Chlorophyceae/ Chaetophorales	Freshwater	France	[26]
Coccomyxa	*olivacea*	**Chlorophyta/** Chlorophyceae/ Chlamydomonadales	Freshwater	France	[26]
Desmococcus	*olivaceus*	**Chlorophyta/** Trebouxiophyceae/ Prasiolales	Freshwater	France	[26]
Fucus	*spiralis*	**Heterokontophyta/** Phaeophyceae/ Fucales	Marine	Fylde coast, UK	[39]
Geminella	*terricola*	**Chlorophyta/** Trebouxiophyceae/ Chlorellales	Freshwater	France	[26]
Halidrys	*siliquosa*	**Ochrophyta/** Phaeophyceae/ Fucales	Marine	Fylde coast, UK	[39]

Table 5.3 contd. ...

...Table 5.3 contd.

Genus	Species	Class/Order/Family	Habitat	Location	References
Hormotila	*mucigena*	**Chlorophyta**/ Chlorophyceae/ Chlorococcales	Freshwater	France	[26]
Keratococcus	*bicaudatus*	**Chlorophyta**/ Trebouxiophyceae/ Chlorellales	Freshwater	France	[26]
Klebsormidium	–	**Charophyta**/ Klebsormidiophyceae/ Klebsormidiales	Freshwater	Dorset, UK/Porto Alegre, Brazil	[44]
	flaccidum			Archamps, France/ Nantes, France/ Laboratory	[3, 21, 42]
	pseudostichococcus			France	[26]
	subtile			Toulouse, France	[40]
Microspora	–	**Chlorophyta**/ Microsporales/ Chlorophyceae	Freshwater	Toulouse, France	[40]
Palmellopsis	*gelatinosa*	**Chlorophyta**/ Chlorophyceae/ Chlamydomonadales	Freshwater	France	[26]
Pleurochrysis	*carterae*	**Prymnesiophyceae**/ Isochrysidales/ Pleurochrysidae	Freshwater	Laboratory	[47]
Pleurococcus	–	**Chlorophyta**/ Chlorophyceae/ Chaetophorales	Freshwater	Dorset, UK/ Carmona, Spain/ Valladolid, Mexico/ Porto Alegre, Brazil	[44]
Polysiphonia	–	**Florideophyceae**/ Ceramiales/ Rhodomelaceae	Marine	Finland	[46]
Sphacelaria	–	**Phaeophyceae**/ Sphacelariales/ Sphacelariaceae	Marine	Finland	[46]
Stichococcus	–	**Chlorophyta**/ Trebouxiophyceae/ Microthamniales	Freshwater	Dorset, UK/ Valladolid, Mexico/ Porto Alegre, Brazil	[44]
	bacillaris			Archamps, France/ Laboratory/ France	[3, 26, 42, 43]

Table 5.3 contd. ...

...Table 5.3 contd.

Genus	Species	Class/Order/Family	Habitat	Location	References
Trebouxia	–	**Chlorophyta/** Trebouxiophyceae/ Microthamniales	Freshwater	France	[26]
Trentepohlia	–	**Chlorophyta/** Ulvophyceae/ Trentepohliales	Freshwater	Dorset, UK/ Carmona, Spain/ Valladolid, Mexico/ Porto Alegre, Brazil/ Toulouse, France/ Portugal	[40, 43, 44]
	iolithus			France	[26, 48]
Ulva	–	**Chlorophyta/** Ulvophyceae/Ulvales	Marine	Fylde coast, UK	[39]
	fasciata		Marine	Puducherry, India	[38]
Porphyra	*umbilicalis*	**Rhodophyta/** Rhodophyceae/ Bangiales	Marine	Fylde coast, UK	[39]
Diatoms					
Unidentified species		–	Marine	Fylde coast, UK	[39]
Aulacoseira	–	**Bacillariophyta/** Coscinodiscophyceae/ Aulacoseirales	Freshwater	Argentina	[45]

pressures is relatively small (between 0.7 and 2.5 MPa) [52]. However, given that a root growing into a concrete crack will be exerting pressure in a manner which will be resolved as a tensile stress in the concrete itself, there still exists the potential for damage.

The tensile strength of concrete is notably low. The following equation has been proposed for the estimation of tensile strength of concrete from compressive cylinder strength [52]:

$$f_t = 0.3 f_c^{\frac{2}{3}}$$

where f_t and f_c are the tensile strength and compressive cylinder strength respectively. Thus, concrete with a cylinder strength of 30 MPa will have a tensile strength of around 2.9 MPa. Furthermore, given that the tips of cracks will also act as sources of stress concentration, the widening of fissures in concrete is less of a challenge for plants than it might immediately seem.

The production of root exudates by plants is likely to exacerbate the magnitude of damage that can be done by a plant, particularly where the substances produced include compounds such as citric acid, whose

Table 5.4 Bryophytes identified growing on concrete surfaces.

Genus	Species	Class/Order	Location	References
Abietinella	*abietina*	**Bryopsida/** Hypnales/ Thuidiaceae	Belarus	[51]
Amblystegium	*serpens*	**Bryopsida/** Hypnales/ Amblystegiaceae	Western Caucasus, Russia/Belarus	[50, 51]
	varium		Western Caucasus, Russia	[50]
Anomodon	*viticulosus*	**Bryopsida/** Hypnales/ Thuidiaceae	Western Caucasus, Russia	[49]
Atrichum	*undulatum*	**Polytrichopsida/** Polytrichales/ Polytrichaceae	Belarus	[51]
Barbula	*unguiculata*	**Bryopsida**/Pottiales/ Pottiaceae	Western Caucasus, Russia/Belarus	[50, 51]
Brachythecium	*albicans*	**Bryopsida/** Hypnales/ Brachytheciaceae	Belarus	[51]
	campestre		Belarus	[51]
	mildeanum		Belarus	[51]
	populeum		Western Caucasus, Russia	[50]
	rivulare		Belarus	[51]
	rotaeanum		Western Caucasus, Russia	[50]
	rutabulum		Western Caucasus, Russia/Belarus	[50, 51]
	salebrosum		Belarus	[51]
	starkei		Belarus	[51]
Brachytheciastrum	*velutinum*	**Bryopsida/** Hypnales/ Brachytheciaceae	Belarus	[51]
Bryoerythrophyllum	*recurvirostrum*	**Bryopsida**/Pottiales/ Pottiaceae	Belarus	[51]

Table 5.4 contd. ...

...Table 5.4 contd.

Genus	Species	Class/Order	Location	References
Bryum	*algovicum*	**Bryopsida**/Bryales/ Bryaceae	Belarus	[51]
	argentium		Western Caucasus, Russia /Belarus	[50, 51]
	bimum		Western Caucasus, Russia	[50]
	caespiticum		Belarus	[51]
	capillare		Belarus	[51]
	creberrimum		Belarus	[51]
	funckii		Lithuania	[49]
	klinggraeffii		Belarus	[51]
	moravicum		Belarus	[51]
	schleicheri		Belarus	[51]
	warneum		Belarus	[51]
Callicladium	*haldanianum*	**Bryopsida**/ Hypnales/ Hypnaceae	Belarus	[51]
Calliergonella	*cuspidata*	**Bryopsida**/ Hypnales/ Hypnaceae	Belarus	[51]
Campyliadelphus	*chrysolphyllum*	**Bryopsida**/ Hypnales/ Amblystegiaceae	Belarus	[51]
Campylidium	*sommerfeltii*	**Bryopsida** / Hypnales / Amblystegiaceae	Belarus	[51]
Campylium	*stellatum*	**Bryopsida**/ Hypnales/ Amblystegiaceae	Belarus	[51]
Ceratodon	*purpureus*	**Bryopsida**/ Dicranales/ Ditrichaceae	Belarus	[51]
Chiloscyphus	*latifolius*	**Jungermanniopsida**/ Jungermanniales/ Lophocoleaceae	Belarus	[51]
	polyanthus		Belarus	[51]
Climacium	*dendroides*	**Bryopsida**/ Hypnales/ Climaciaceae	Belarus	[51]

Table 5.4 contd. ...

...Table 5.4 contd.

Genus	Species	Class/Order	Location	References
Cololejeunea	*rossetiana*	**Jungermanniopsida/** Porellales/ Lejeuneaceae	Western Caucasus, Russia	[50]
Conocephalum	*conicum*	**Marchantiopsida/** Marchantiales/ Conocephalaceae	Belarus	[51]
Ctenidium	*molluscum*	**Bryidae**/Hypnales/ Hylocomiaceae	Western Caucasus, Russia	[50]
Dicranella	*cerviculata*	**Bryopsida/** Dicranales/ Dicranellaceae	Belarus	[51]
	heteromalla		Belarus	[51]
Dicranum	*flagellare*	**Bryopsida/** Dicranales/ Dicranaceae	Belarus	[51]
	scoparium		Belarus	[51]
Didymon	*rigidulus*	**Bryopsida**/Pottiales/ Pottiaceae	Belarus	[51]
Dreponacladus	*polycarpus*	**Bryopsida/** Hypnales/ Amblystegiaceae	Belarus	[51]
Encalypta	*streptocapra*	**Bryopsida/** Encalyptales/ Encalyptaceae	Belarus	[51]
Eurhynchium	*angustirete*	**Bryopsida/** Hypnales/ Brachytheciaceae	Belarus	[51]
	crassinervium		Western Caucasus, Russia	[50]
Fissidens	*adianthoides*	**Bryopsida/** Dicranales/ Fissidentaceae	Belarus	[51]
	dubius		Lithuania	[49]
Fontinalis	-	**Bryopsida/** Hypnales/ Fontinalaceae	Argentina	[45]
Funaria	*hygrometrica*	**Bryopsida**/Funariales/ Funariaceae	Belarus	[51]
Grimmia	*muehlenbeckii*	**Bryopsida/** Grimmiales/ Grimmiaceae	Belarus	[51]
	pulvinata		Western Caucasus, Russia	[50]
Gymnostomum	*aeruginosum*	**Bryopsida**/Pottiales/ Pottiaceae	Western Caucasus, Russia	[50]
Hedwigia	*ciliata*	**Bryopsida/** Leucodontales/ Hedwigiaceae	Belarus	[51]

Table 5.4 contd. ...

...Table 5.4 contd.

Genus	Species	Class/Order	Location	References
Homalia	*trichomanoides*	**Bryopsida**/ Hypnales/ Neckeraceae	Belarus	[51]
Homalothecium	*lutescens*	**Bryopsida**/ Hypnales/ Brachytheciaceae	Belarus	[51]
	sericeum		Western Caucasus, Russia	[50]
Homomallium	*incurvatum*	**Bryopsida**/ Hypnales/Hypnaceae	Lithuania	[49]
Hygroamblystegium	*juratzkanum*	**Bryopsida**/ Hypnales/ Amblystegiaceae	Belarus	[51]
	varium		Belarus	[51]
Hygrohypnum	*luridum*	**Bryopsida**/Hypnales/ Campyliaceae	Western Caucasus, Russia	[50]
Hylocomium	*splendens*	**Bryopsida**/Hypnales/ Hylocomiaceae	Belarus	[51]
Hypnum	*cupressiforme*	**Bryopsida**/ Hypnales/Hypnaceae	Belarus	[51]
Jungermannia	*atrovirens*	**Jungermanniopsida**/ Jungermanniales/ Jungermanniaceae	Western Caucasus, Russia	[50]
Leskea	*polycarpa*	**Bryopsida**/ Hypnales/Leskeaceae	Belarus	[51]
Leptobryum	*pyriforme*	**Bryopsida**/Bryales/ Bryaceae	Belarus	[51]
Leucodon	*sciuroides*	**Bryopsida**/ Leucodontales/ Leucodontaceae	Belarus	[51]
Marchantia	*polymorpha*	**Marchantiopsida**/ Marchantiales/ Marchantiaceae	Belarus	[51]
Mnium	*marginatum*	**Bryopsida**/Bryales/ Mniaceae	Belarus	[51]
Orthotrichum	*anomalum*	**Bryopsida**/ Orthotrichales/ Orthotrichaceae	Belarus	[51]
	cupulatum		Belarus	[51]
	diaphanum		Belarus	[51]
	gymnostomum		Belarus	[51]
	obtusifolium		Belarus	[51]
	pallens		Belarus	[51]
	patens		Belarus	[51]
	pumilum		Belarus	[51]
	speciosum		Belarus	[51]

Table 5.4 contd. ...

...Table 5.4 contd.

Genus	Species	Class/Order	Location	References
Oxyrrhynchium	*hians*	**Bryopsida/** Hypnales/ Brachytheciaceae	Western Caucasus, Russia/Belarus	[50, 51]
Oxystegus	*tenuirostris*	**Bryopsida**/Pottiales/ Pottiaceae	Western Caucasus, Russia	[50]
Platyhypnidium	*riparioides*	**Bryopsida/** Hypnales/ Brachytheciaceae	Western Caucasus, Russia	[50]
Pohlia	*nutans*	**Bryopsida**/Bryales/ Mniaceae	Belarus	[51]
Pseudoleskeella	*catenulata*	**Bryopsida/** Hypnales/ Pseudoleskeellaceae	Lithuania	[49]
	nervosa		Belarus	[51]
Plagiochila	*porelloides*	**Jungermanniopsida/** Jungermanniales/ Plagiochilaceae	Belarus	[51]
Plagiomnium	*affine*	**Bryopsida**/Bryales/ Mniaceae	Belarus	[51]
	cuspidatum		Belarus	[51]
	elatum		Belarus	[51]
	ellipticum		Belarus	[51]
	rostratum		Western Caucasus, Russia	[50]
	undulatum		Belarus	[51]
Plagiothecium	*laetum*	**Jungermanniopsida/** Jungermanniales/ Plagiochilaceae	Belarus	[51]
Pleurozium	*schreberi*	**Bryopsida**/Hypnales/ Hylocomiaceae	Belarus	[51]
Polytrichum	*formosum*	**Polytrichopsida/** Polytrichales/ Polytrichaceae	Belarus	[51]
	juniperinum		Belarus	[51]
Pylaisia	*polyantha*	**Bryopsida/** Hypnales/ Pylaisiaceae	Western Caucasus, Russia/Belarus	[50, 51]
Racomitrium	*canescens*	**Bryopsida/** Grimmiales/ Grimmiaceae	Belarus	[51]
Rhynchostegiella	*teneriffae*	**Bryopsida/** Hypnales/ Brachytheciaceae	Western Caucasus, Russia	[50]

Table 5.4 contd. ...

...Table 5.4 contd.

Genus	Species	Class/Order	Location	References
Rhynchostegium	*confertum*	**Bryopsida/** Hypnales/ Brachytheciaceae	Western Caucasus, Russia/Belarus	[50, 51]
Rhytidiadelphus	*squarrosus*	**Bryopsida/** Hypnales/	Belarus	[51]
	triquetrus	Hylocomiaceae	Belarus	[51]
Sanionia	*uncinata*	**Bryopsida**/Hypnales/ Scorpidiaceae	Belarus	[51]
Sciuro-hypnum	*oedipodium*	**Bryopsida/** Hypnales/	Belarus	[51]
	populeum	Brachytheciaceae	Belarus	[51]
Serpoleskea	*subtilis*	**Bryopsida/** Hypnales/ Amblystegiaceae	Belarus	[51]
Stereodon	*fertilis*	**Bryopsida/** Hypnales/	Belarus	[51]
	pallescens	Hypnaceae	Belarus	[51]
Schistidium	*apocaprum*	**Bryopsida/** Grimmiales/	Belarus	[51]
	crassipilum	Grimmiaceae	Belarus	[51]
Syntrichia	*papillosa*	**Bryopsida**/Pottiales/ Pottiaceae	Lithuania	[49]
	ruralis		Belarus	[51]
	virescens		Belarus	[51]
Thuidium	*assimile*	**Bryopsida**/Hypnales/ Thuidiaceae	Belarus	[51]
Tortella	*tortuosa*	**Bryopsida**/Pottiales/ Pottiaceae	Belarus	[51]
Tortula	–	**Bryopsida**/Pottiales/ Pottiaceae	Portugal	[43]
	mucronifolia		Belarus	[51]
	muralis		Belarus	[51]
	obtusifolia		Lithuania	[49]
Weissia	*contraversa*	**Bryopsida**/Pottiales/ Pottiaceae	Belarus	[51]

damaging effects have already been discussed in Chapter 4. Research into this aspect of deterioration currently appears to be non-existent. Most of the other acids produced by plant roots have been dealt with in previous chapters. However, formic and fumaric acid have not. Examination of the nature of the interaction between these acids and calcium, iron and aluminium ions in Chapter 2 indicates that their effect will largely be that of acidolysis.

A special case with regards to interaction of vascular plants with concrete structures is that of climbing plants (Figure 5.10). There has been a great deal of debate with regards to whether such plants are harmful to the walls on which they grow, with particular attention to the ivy, *Hedera helix*. One study of ivy growing on a specially constructed stone test wall found that the plant was not particularly predisposed to enter holes and cracks [54]. However, observation of ivy on historic structures has identified many instances where the roots enter existing voids in walls leading to damage [54, 55]. It has been proposed that this process is promoted by shortages of water, and that events such as the cutting off of the stem of the plant at the base may cause smaller rootlets to permeate such voids more aggressively to obtain water [54].

The types of voids which are likely to be amenable to plants such as ivy are likely to be far scarcer on a concrete structure, compared to a stone wall, but might include joints, or cracks resulting from other forms of deterioration. Where the incursion of roots into the fabric of a structure is unlikely, there is some evidence that ivy provides protection to buildings, principally in its ability to capture air-borne particulates before they become attached to the building surface. It is argued that this has the effect of keeping the surfaces themselves clean, which may also have the additional benefit of limiting the delivery of nutrients to the surface which might otherwise be used by micro-organisms in biodeterioration processes [56].

Figure 5.10 A climbing plant on a concrete wall.

Other climbing plants attach themselves to vertical surfaces using suckers. These suckers sometimes exude organic acids. For instance, the Virginia creeper (*Parthenocissus quinquefolia*) exudes oxalic acid from its suckers and other parts [57]. It has been suggested that exudation of acid by suckers can potentially damage the underlying wall [83]. However, doubts have been expressed, certainly with regards to whether oxalic acid production will lead to deterioration [57].

Another issue related to root growth is the damage to concrete pavements, foundations and pipes by tree roots. The growth of roots is generally directed downwards and away from sunlight (negative phototropism) [58]. However, the direction of growth is also influenced by concentration gradients of water, nutrients and oxygen. Where areas around trees are paved with concrete, or other surfacing materials, the highly obscuration of sunlight is unlikely to suppress upward growth. Moreover, impermeable paving materials tend to prevent the evaporation of water from soil, leading to condensation of water at the soil–concrete interface. Thus, root growth very close to this interface is not uncommon. The problem is often made worse by the presence of highly compacted soil beneath a pavement which resists root penetration and contains limit concentrations of oxygen. Subsequent radial growth of the roots as the tree matures can then lead to the generation of flexural stress directly underneath the concrete, producing cracking and buckling.

Another mechanism by which tree roots can damage structures is through the uptake of water from soils prone to shrinkage. Whether a soil can undergo significant shrinkage is dependent on whether expansive clays (such as montmorillonites) are present, the size of the clay particles (with smaller particle sizes yielding greater shrinkage) and the extent to which clay particles are aligned in a preferred orientation within the soil (with a poor degree of alignment leading to larger quantities of shrinkage [58]).

Where trees are present in such soils, their uptake of water will lead to accelerated drying during dry periods, which is liable to lead to an amplification in the magnitude of shrinkage. This can present problems for the foundations of nearby low-rise buildings [59] since movement of the soil can generate stresses in foundations which can cause substantial damage.

The extent to which trees threaten foundations is partly dependent on the water demand of the species of tree. Trees with particularly high water demand include elm, oak, poplar, willow and cypress species [60].

It is usually convenient, for a number of reasons, for the movement of water across distances (either as drinking water, sewage or drainage water) to be conducted via pipes located underground. However, their need to locate water means that the roots of plants will grow towards underground pipes and through the mechanisms already discussed, invade pipes, usually through joints. This has two detrimental effects. Firstly, once breached, the pipe will leak, with environmental and efficiency implications. Secondly, the pipe will gradually become blocked by the invading root (see Figure 5.11).

A number of studies have been conducted into this phenomenon, with many taking the form of surveys of water networks in specific geographical locations. Such studies have been made possible through the use of mobile cameras that can be run through pipes to inspect their interior. One such study, conducted in Sweden, examined both the frequency of intrusion and related each instance of intrusion to tree and shrub species growing in close proximity to the pipes [62]. The species found to be most frequently responsible for intrusions were *Malus floribunda* (Japanese flowering crabapple) and *Populus candensis* (Canadian poplar). The two main types of pipe encountered in the survey were concrete and PVC, and it is interesting to note that the concrete pipes were considerably more resistant to intrusion than the polymer pipes. The survey detected 0.661 root intrusions per pipe joint in the case of PVC, whilst the frequency was only 0.080 in the case of concrete. This may reflect the greater mass of the concrete pipes which would be likely to present greater resistance to the growth pressures of roots.

A similar study conducted in Poland made a similar observation, although in this case the two types of pipe encountered were concrete and vitrified clay [61]. The frequency of intrusion was expressed in a slightly different form (intrusions per 100 m). Concrete pipes proved more resistant, with an intrusion rate of 3.24 intrusions per 100 m, compared to a value of 3.99 for vitrified clay pipes. Smaller diameter pipes are also more vulnerable to intrusion [63].

Figure 5.11 Root intrusion in a pipe [61].

A survey of pipes in Melbourne, Australia found that the two tree species most likely to be involved in root intrusion were of the genera *Eucalyptus* and *Melaleuca* [64].

Reports of the deterioration of concrete surfaces resulting from the establishment of moss are scarce and based purely on observation. Where this does appear to have occurred, it has been attributed to the penetration of rhizoids into the material. A study of lichen and moss on cathedrals in Spain found that bryophytes growing on the mortar of stonework appeared to be responsible for its disintegration in this way [65]. It should be noted, however, that penetration of rhizoids may have occurred through cracks resulting from prior deterioration.

5.5 Deterioration from Algae

The issue of whether the attack of concrete by algae is a real phenomenon is still to be decided. It is certainly true that the cement matrix—and possibly the aggregate—of concrete contain nutrients for algae, and the uptake of these nutrients will lead to the development of porosity in the material [66]. Moreover, the production by algae of organic acids likely to accelerate this process is confirmed. However, what is theoretically possible is only partly supported by scientific evidence.

Whilst a few of studies have examined the interaction of algae with concrete, with the assumption that the interaction is one of biodeterioration, this has typically not included quantitative measurements of loss of mechanical properties or mass. One group of researchers have identified chemical changes in concrete surfaces resulting from colonisation by marine algae–probably *Ulva fasciata* [38, 67]. The results of energy dispersive X-ray spectroscopy and powder X-ray diffraction analysis certainly indicate a change in the composition of colonised concrete surfaces relative to similar surfaces exposed to only water. Specifically, there appears to be an increase in the quantity of calcium at the concrete surface at the expense of other elements. Whether this is the result of removal of material or the precipitation of calcium-rich material at the surface is not clear, although the presence of higher concentrations of calcite at the colonised surfaces points to the latter. The influence of these changes on concrete properties was not investigated further.

Another study has employed microscopy as a means of observing the manner in which marine algae interact with concrete armour deployed in coastal protection applications [39]. The study revealed the presence of the algae species *Ascophylum nodosum, Fucus vesiculosus, Fucus spiralis, Porphyra umbilicalis, Halidrys siliquosa* and species of the genus *Ulva*. Microscopy revealed the growth of algal filaments in the interface between fine aggregate and the cement matrix, with the filaments also growing through the cement matrix in places. The researchers propose that this process of growth may contribute towards the erosion of such materials, alongside the abiotic processes acting on coastal protection structures in a coastal environment. They also propose that the use of marine aggregate already carrying algae may introduce these organisms into concrete, with the algae surviving the initial high pH conditions in concrete as dormant forms such as spores. Both hypotheses are credible, but evidence of either is very limited.

It has also been argued that the presence of algae may also protect concrete surfaces from the mechanical action of the marine environment which can cause deterioration [68].

Research has examined the interaction of marine algae with macro- and micro-synthetic fibres used within similar concrete components. In this case, deterioration was more conclusively demonstrated [69, 70]. The macro fibres included were composed of both polyethylene and polypropylene, whilst the microfibers were polyethylene. Colonising the concrete surface was algae from the genus *Ulva*. Microscopy identified numerous instances where the algal filaments had grown into the fibres leading to a loss of cross-sectional area. It has been demonstrated that algae can use polyethylene as a substrate, although whether they can utilise carbon in the polymer is unclear [71]. The researchers detail a proposed mechanism by which deterioration of fibres may ultimately lead to more substantial loss of mass from the concrete as a whole.

Algae has also been blamed for accelerated corrosion of steel reinforcement in concrete [66]. The proposed mechanism for this process centres around the photosynthetic nature of algae. During periods of daylight, algae will undergo photosynthesis, producing oxygen. Thus, the presence of algae at the surface of a reinforced concrete structural element may produce localised oxygen enrichment. Such enrichment has the potential to create an 'aeration cell' in which two locations around a steel reinforcement bar experience very different oxygen concentrations. This can lead to an electrochemical cell being established with localised corrosion producing pitting of the steel.

There is no question that microbially induced corrosion of exposed metal surfaces is a real phenomenon [72]. However, whether it is a mechanism that can lead to corrosion of steel embedded in reinforced concrete is less certain. The same researchers who propose this mechanism have conducted a preliminary investigation comparing the corrosion potential of steel embedded in concrete cylinders (with a cover depth of 20 mm) and submerged in a bioreactor illuminated with a light source operating under a 12 hours on—12 hours off regime. The bioreactor contained a culture medium in which the algae *Pleurochrysis carterae* had been cultured.

Two types of cement were evaluated—Portland cement and a geopolymer cement made using fly ash. In the case of the Portland cement concrete, the corrosion potential of the steel maintained a very slightly negative value (indicative of a low potential for corrosion) throughout the 14 day experimental period. Evaluation of cell numbers showed that the algae were completely eradicated by 7 days, presumably as a result of the high pH of the culture medium (around 10) originating from the Portland cement. Where the geopolymer concrete was exposed to the culture medium in the absence of algae, the corrosion potential rapidly dropped to a more negative value, indicating a greater potential for corrosion. This was presumably either due to the lower pH of the cement limiting passivation of the steel surface or the higher porosity of the cement matrix, or a combination of both. However, when algae were present the corrosion potential remained at a level comparable to the Portland cement concrete until the end of the 14-day experiment, at which point there was an abrupt drop to a more negative value. This drop approximately coincided with the extinction of the algae population, leading the researchers to propose that the algae might, in fact, be protecting the steel by maintaining a higher pH through oxygen production.

Regardless of the precise nature of the processes occurring in these experiments, there would appear to be no reported examples of such a process occurring in the field. The main argument against such a process is that, for it to be effective, there would need to be considerable diffusion of oxygen into the concrete towards the steel surface. This is because algal activity will be limited to a zone very close to the concrete surface, due to the need for light. Because of the uncertainties surrounding this proposed form of biodeterioration, it is not explored further in this chapter.

Nonetheless the presence of algae on concrete surfaces is often prominently visible and the resulting stain can be detrimental in aesthetic terms, particularly on lighter surfaces. Moreover, the presence of algae on horizontal surfaces on which people walk can be slippery and consequently hazardous.

5.6 Limiting Biodeterioration from Plants

It is hopefully evident that algal growth on concrete surfaces and the establishment and growth of vascular plants on concrete structures are very different processes. For this reason, it is necessary to distinguish between the measures which can be used against algae and vascular plants.

5.6.1 Limiting algal growth

Approaches to controlling algal growth on concrete can be categorised in terms of whether they involve control of environmental conditions, concrete composition or treatments applied to the hardened concrete surface. These different approaches are discussed separately below.

Environmental factors

Aside from the nutrient needs of algae, the organisms require water and sunlight. Thus, where either of these are absent, algae will be able to live. Thus, surfaces which are permanently in darkness cannot be occupied. Figure 5.12 shows the generation time (the time required for the population to double) of a species of algae. At low levels of light intensity, very long generation times are observed. However, as light intensity increases—for this particular case—the generation time increases again beyond the optimum intensity of around 60 W/m². The intensity of sunlight incident on the Earth's surface will depend on the latitude and time of year (with greater seasonal variation for latitudes further away from the equator). However, intensities will typically be considerably higher than 60 W/m² at noon on a clear day (for instance around 500 W/m² in spring and autumn in the South of England). However, this is unlikely to compromise algal growth too much, and many surfaces will seldom experience the full intensity of sunlight where surfaces are horizontal, or where there is partial shading and cloud cover.

Parts of a concrete structure which experience low levels of moisture are unlikely to support algal growth. The environment surrounding a structure will also define the availability of moisture. A survey involving statistical analysis of algal colonisation of building façades across France found that relative humidity was the most important factor in influencing algal growth, with higher quantities of precipitation and closer proximity of vegetation also encouraging growth [74].

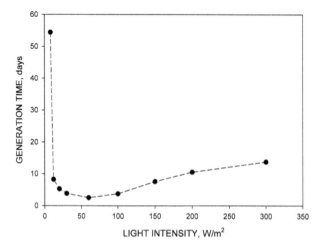

Figure 5.12 The generation time of the algae *Botryococcus braunii* exposed to different intensities of light in fluid media held in flasks. Light source not specified [73].

Thus, interiors and exterior surfaces which receive shelter from rainfall from other parts of a structure (such as cantilevered projections, awnings and roof eaves) will potentially remain sufficiently dry to resist the establishment of algae. Indeed, in designing a structure, there may exist possibilities for incorporating features which will limit the wetting of concrete surfaces, including the use of coping and capping, or projecting sills.

Algae were found to favour the North- and West-facing facades of buildings in France [74]. This was attributed to the North side of buildings receiving less sun, and consequently drying out less frequently. Favourable colonisation of West-facing surfaces was considered the result of dominant westerly winds bringing rain from the Atlantic Ocean. Such patterns will clearly vary according to the geographic location of a building, and knowledge of local conditions will at least allow identification of which parts of a structure require measures to control growth.

Temperature ranges over which algae will grow vary between species. However, for most species optimum growth rates occur somewhere between 20°C and 35°C, although some algae continue to grow at higher temperatures. This includes *Chlorella vulgaris*, which can grow on concrete surfaces (Table 5.3) and which is capable of growth at temperatures as high as 40°C [75].

These ranges are clearly widely experienced as ambient temperatures, meaning that the vast majority of concrete structures worldwide will offer suitable environments for algae, at least at certain parts of the year. Moreover, the control of external temperatures is not a practical option.

The amplitude of temperatures throughout the year has also been found to play an important role, with smaller magnitudes of fluctuation limiting evaporation rates and maintaining levels of moisture [74].

Material characteristics

As discussed in Section 5.3.4, the small volume of porosity resulting from a low water/cement ratio will limit the extent to which the cement matrix of concrete will act as a reservoir of moisture for algae. This is also likely to limit the extent to which attachment of algae can occur, since a lower water/cement ratio is also likely to yield a smoother surface. Cement type will have little influence on rates of growth in the long term: initially Portland cement will have a pH higher than that of combinations of Portland cement and other materials, or calcium aluminate cements. However, the alkalinity of the surface will gradually decline as the surface is leached by water or undergoes carbonation.

Limiting algal growth through the use of TiO_2 in the form of the mineral anatase—in the same manner as for fungi in Chapter 4—has been demonstrated. Research investigating the mineral's effectiveness against algal growth used the material as an addition in mortars which were subsequently inoculated with *Stichococcus bacillaris* and *Chlorella ellipsoidea*, plus the cyanobacterium *Gleocapsa dermochroa* [43]. Mortar was also made containing iron-doped anatase. The logic behind this approach was that anatase doped in this manner had previously been found to possess a higher photocatalytic efficiency [76]. The mortar composition was 12:4:4:1 sand:lime:anatase:Portland cement. Additionally, other mortar surfaces were treated with two different biocides: alkyl-benzyl-dimethyl-ammonium chloride/isopropyl alcohol (commercial name: Biotin T) and n.n-didecyl-n-methyl-poly(oxyethyl) ammonium propionate/alkyl-propylene-diamineguanidium acetate (Anios DDSH). Measurement of algal growth over a 4 month period found that the anatase-bearing mortar was the most resistant, followed by Biotin T (Figure 5.13). Interestingly, the doped anatase was less effective.

A similar study also found that a commercial white cement containing an unspecified quantity of TiO_2 was wholly effective in preventing algal growth (*Chlorella vulgaris*) over an experimental period of 16 weeks [77]. Similar laboratory-made materials consisting of reagent grade TiO_2—at 5% and 10% by mass—in white cement also containing GGBS were ineffective. It was proposed that the mixes might have been more porous, encouraging algal attachment and growth.

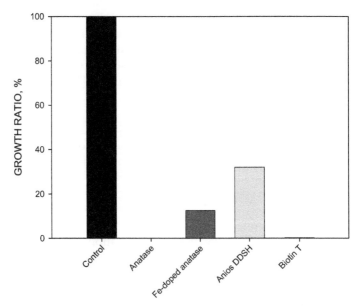

Figure 5.13 Growth rate of algae and cyanobacteria on mortar surfaces, expressed as the quantity of chlorophyll a extracted as a percentage of the control [43].

Protection after construction

A study examining the effectiveness of different façade coatings measured their ability to support growth of three different species of algae— *Klebsormidium flaccidum, Chlorella mirabilis* and *Stichococcus bacillaris* [42]. The coatings investigated were three mortars, an acrylic water-based organic coating and a similarly formulated paint. Neither the paint or the organic coating contained a biocide. The paint was considerably better at limiting growth in comparison to the other materials (Figure 5.14). Indeed, a study characterising and quantifying algal colonisation of a large number of building façades in France found that concrete and mortar surfaces were more amenable to the establishment of algal populations than organic surfaces such as paint [74], since the isolation of algae from the porosity of the concrete limits the availability of water.

The use of water repellents and biocides against algae growth (in this case *Chlorella vulgaris*) on white architectural concrete and precast aerated cellular concrete has been investigated [41]. In the case of the white concrete, both biocides and water repellents were highly effective for the full test period of 12 weeks. The treatments were less effective on aerated cellular concrete, possibly as a result of this material's more porous nature. It is interesting to note that different water repellents had very different effects on algal growth on aerated concrete, with some actually accelerating growth. This is possibly explained by the enhanced ability for some algae to attach themselves to hydrophobic surfaces (see Section 5.3.1). However, other

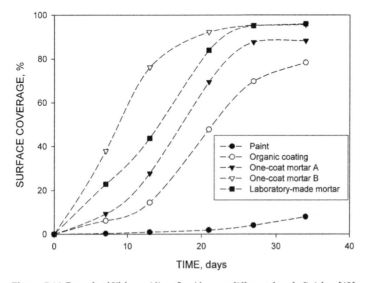

Figure 5.14 Growth of *Klebsormidium flaccidum* on different façade finishes [42].

water repellents were more effective—specifically a silane-based product. Biocides tended to prolong the period of time before algal growth started, but growth occurred at a comparable rate to controls once this period was over. The most effective strategy was found to be the combination of water repellent and biocide—a chlorinated pyridine-based formulation (Figure 5.15). Regardless of this, it is evident from the results of this study that both approaches are likely to provide protection from algal growth for a finite period. It should also be noted that the use of biocides on concrete surfaces may be unacceptable in some applications—for instance, in coastal or marine applications.

Whilst is has been seen that anatase additions to concrete and mortar can limit algal growth, anatase-bearing coatings applied after construction or precast manufacture of concrete could provide more economic protection, with a greater permanency than biocides. One study has examined the ability of TiO_2 applied to the surface of autoclaved aerated concrete to control algal growth (*Chlorella vulgaris*) [77]. The TiO_2 was applied using a vacuum impregnation technique. However, algae growth on this coated surface appeared to be largely unaffected compared to the untreated control surface. Indeed, a silane-based water repellent treatment proved considerably more effective.

Nonetheless, more conventional coatings of TiO_2 particles in a PMMA binder (mass ratio 9:1) applied to a cement paste substrate have been shown to be effective [78]. The use of a tungsten oxide (WO_3) coatings was also investigated, but proved ineffective. However, the approach was further

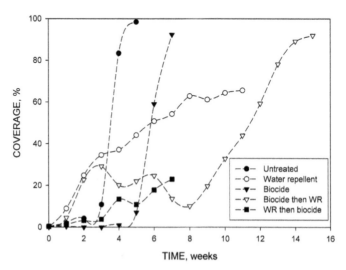

Figure 5.15 Development of algal coverage of autoclaved aerated concrete with surface treatments of water repellent, biocide and combinations of both where treatments were applied in different orders [41].

refined through the use of both of these oxides coated with co-catalysts of either platinum (Pt) or iridium (Ir) which proved to be highly effective at limiting growth. It was demonstrated that the presence of these elements was effective through a photocatalytic mechanism rather than one of toxicity through limiting the light source used to wavelengths that did not contribute significantly to the photocatalytic effect, which yielded growth at a greater rate. Whether such an approach is practical for concrete structures is questionable: both metals are extremely expensive.

The conclusion that must be drawn from these findings is that anatase and the application of paint appear to offer the best means of achieving concrete surfaces which limit algal growth. However, the use of water repellent coatings, should not be entirely ruled out. In all cases, surface coatings must be viewed as being non-permanent, and requiring re-application later in a structure's life.

Cleaning

Removal of algae from concrete surfaces can be approached in a number of ways, but the main decision which needs to be made is whether a biocide can or should be used. Biocides will kill the occupying algae, but normally the dead cells will remain attached to the surface [79]. Removal can be achieved through scraping or brushing, and in extreme cases abrasive blasting can be used. This last approach, whilst highly effective at removing the residue will also abrade the surface, which may not be desirable. Removal can also be achieved through the use of high velocity water jets, which are highly effective and less likely to damage the underlying concrete.

The duration of a biocide treatment is unlikely to be more than one or two years [80]. Moreover, given that high velocity jetting will usually remove algae regardless of whether a biocide has been used, there exists a good argument for using this technique on its own.

The removal of stains resulting from algae using bacteria has been investigated in the laboratory [81]. The experimental programme involved concrete cubes which had been exposed to the elements for a period of several years and consequently developed a covering of algae. The cubes were placed in a volume of water such that only the upper face of each cube was not in contact with water, to ensure that this surface remained moist. The upper surface of the cubes was then sprinkled with droplets of a solution consisting of water, bacteria of the genus *Thiobacillus* (see Chapter 3), powdered sulphur and ammonium sulphate. Sprinkling was conducted four times a day for three days. After this period of treatment, the cubes were dried and the treatment repeated a further two times.

The logic behind this approach is that the bacteria will utilise the sulphur in the solution as a source of energy, producing sulphuric acid. This acid will then react with the concrete surface, leading to a cleaning effect.

The change in colour of the surface during this treatment was measured using colorimetry. The treatment was found to yield a change in colour that indicated a cleaning effect. This was not as effective as submerging the cubes in the solution, or, indeed, submerging the cubes in a solution of sulphuric acid. Petrographic microscopy of the surface indicated a layer of gypsum has been formed (see Chapter 3). It is sound to assume that this layer would eventually be weathered leaving a wholly clean surface behind. However, whilst such a process would appear to be effective, the use of sulphuric acid—regardless of its source—as a means of cleaning will clearly lead to a loss of mass from the surface and might be deemed overly aggressive.

5.6.2 Limiting damage from vascular plants

As already discussed, the establishment of plants on a concrete surface will usually either require the presence of cracks or open joints in which seeds can become lodged, or a horizontal layer of moss in which particle can become trapped. Thus, part of a strategy of preventing of plants from becoming established should be in limiting the occurrence of such features.

Cracks can arise in concrete for a wide range of reasons at various points in the life of a structure. Causes of cracking can include the thermal contraction of concrete shortly after placement, drying of concrete in the fresh state (plastic shrinkage), drying of concrete elements in the hardened state under restraint (drying shrinkage), freeze-thaw attack, alkali-silica reaction, sulphate attack, and the corrosion of steel reinforcement [82]. Detailing suitable approaches to designing concrete mixes to prevent these forms of deterioration is beyond the scope of this book, but clearly adequate design of concrete for durability is a desirable prerequisite for structures which are to resist plant colonisation.

Where cracks are already present, they will need to be repaired, not only to prevent plant establishment, but to prevent other forms of deterioration, and to satisfy serviceability requirements of the structure. The repair of cracks will normally involve either the use of structural repair materials, which are applied by hand, or the use of concrete injection methods. Structural repair materials are normally either inorganic cements or polymer modified cements. Injection formulations are typically based on polyurethane or epoxy resins.

Joints are a necessary feature in many parts of concrete structures, since they prevent cracking resulting from volume change. Moreover, discontinuities between different structural elements are often an inevitable aspect of the construction process. However, they clearly offer an opening in the concrete surface in which plants can be established (Figure 5.16). Therefore, sealing of joints is a prudent strategy. Indeed, there are more general reasons for sealing joints, since they also present a means for substances which are likely to compromise the durability of concrete to make their way beneath the surface. Sealants used for joints are normally highly elastic materials which are impermeable to water. Normally it is anticipated that sealants will begin to deteriorate within the intended service life of a concrete structure, and so replacement at some point is usually inevitable. Thus, seals should be inspected as part of the maintenance programme of a structure.

Where precast elements are placed alongside each other, a joint is created which can also accomodate plants. An interesting example of this is reported in the form of the penetration of Japanese knotweed (*Fallopia japonica*) through precast concrete river revetment panels [18]. This problem was ultimately solved by fabricating the panels with interlocking edges which resisted penetration.

The presence of moss on concrete surfaces is, like lichen, quite often aesthetically appealing. However, its potential role as a mechanism by which a growth medium for vascular plants can be established, means that it may not always be desirable. It should be noted that for vascular plants

Figure 5.16 Plant growing in a joint in a concrete structure.

to damage the underlying concrete, cracks or joints will still probably need to be present.

Removal of moss is most effectively achieved through physical methods such as scraping or water jetting—removal is the important part, and application of herbicide will not make this process any easier.

The design of concrete mixes to resist damage from plant damage can be done with less certainty compared to the organisms examined in previous chapters. A reduced water/cement ratio is certainly unlikely to do any harm, could potential reduce the availability of water, and would have the side-effect of improving concrete durability in more general terms. Furthermore, assuming that the worst-case scenario for acid exudation from roots is citric acid, the results reported in Chapter 4 would point to either Portland or calcium aluminate cement.

Where vascular plants have already become established on a structure, their removal is necessary, followed by filling of cracks with appropriate repair materials, and joints with sealants, to prevent re-establishment. In some cases the application of herbicides may be beneficial.

Herbicides can be categorised in terms of the manner in which they work and can be selected accordingly. *Hormone* type herbicides mimic the growth hormones of plants. They selectively target broad-leafed plants, but do not harm narrow-leafed plants, such as grasses.

Moss and algae herbicides are particularly suited for acting on these plants. They include acetic acid and ferrous sulphate, both of which are not suited for use on concrete. Ferrous sulphate will stain concrete, whilst the potentially damaging effect of acetic acid on concrete is covered in Chapter 3. However, acetic acid products marketed as both algae/moss herbicides and cleaning agents are sometimes marketed specifically for cleaning concrete, on the grounds that the acid will remove the outer surface of the paving as part of the cleaning process.

Contact herbicides kill only the part of the plant that they come into contact with, whilst *systemic* herbicides will move through the plant to affect all parts. Residual herbicides take some time either to be dispersed or break down, thus leaving soil uninhabitable by plants for some time after application.

Table 5.5 lists herbicides which are likely to potentially be usable on concrete. It should be noted that most commercial products are combinations of more than one herbicide, to tailor the product to specific applications.

Climbing plants

The establishment of climbing plants on concrete walls is, again, possibly an agreeable development, and the potential for damage should be carefully

Table 5.5 Herbicides likely to be suitable for use on concrete.

Herbicide	Residual?	Type	Suitable For
Triclopyr	Slightly	Hormone	Broad-leafed plants
2,4-D	Slightly	Hormone	Broad-leafed plants
Dicamba	Slightly	Hormone	Broad-leafed plants
Mecoprop-P	Slightly	Hormone	Broad-leafed plants
MCPA	Slightly	Hormone	Broad-leafed plants
Clopyralid	Slightly	Hormone	Broad-leafed plants
Fluroxypyr	Slightly	Hormone	Broad-leafed plants
Benzalconium chloride	No	Moss and algae herbicides	Moss, algae
Fatty acids (including pelargonic acid)	No	Moss and algae herbicides + Contact	Moss, algae, vascular plants
Diquat	No	Contact	Vascular plants
Glyphosate	No	Systemic	Vascular plants
Flufenacet	Yes	Residual	Vascular plants
Metosulam	Yes	Residual	Vascular plants
Diflufenican	Yes	Residual	Vascular plants

considered prior to deciding to remove such plants. This is because concrete walls are likely to possess fewer features which would allow roots or tendrils to intrude beneath the surface, and where these are absent, there is little argument for removal.

Where the plant is ivy, and damage is a real possibility, guidance has been developed for removal [83]. The approach recommended takes into account the fact that the adventitious roots of ivy may continue to grow—possibly more aggressively—after the main stem has been cut. The advised sequence of events is as follows:

1. The main stem of the plant is cut.
2. Where the plant is well established, it should be sprayed with a systemic herbicide.
3. The plant should be left for a period of time until the stems have died and show evidence of shrinkage.
4. The roots should either be dug up, or an herbicidal gel applied to the stump.
5. The plant should be removed from every point of entry into the wall, and the wall repaired if necessary.

Tree roots and pavements

The extent to which roots are able to grow under pavements can be limited by the use of root barriers. Root barriers are layers of material—running vertically between the tree and the pavement to be protected—whose purpose is to deflect or constrict the growth of roots [58]. Materials able to deflect growth include plastic and glass fibre composite panels, preserved plywood and tar paper or asphalt impregnated felt. Constricting barriers are normally geotextiles, but can also include copper mesh. This material possibly has an enhanced control over roots, in that copper released into the soil is likely to have an inhibitory effect on growth. Indeed, geotextiles impregnated with copper sulphate have also been successfully used as a root barrier.

It has been argued that the use of root barriers compromises the stability of trees, but a study simulating the effect of wind on trees grown with and without barriers found that the presence of barriers actually appeared to increase resistance to being uprooted [84]. This was attributed to the deeper roots of these trees.

Another possible approach to limiting upward growth of roots beneath pavements is to allow moisture in soil to evaporate, leading to lower moisture levels at the pavement–soil interface. This can be achieved through permeable paving. Where the paving is concrete, this can take the form either of paviers with gaps which allow the movement of water vapour—but still support foot traffic—or pervious concrete. Pervious concrete is concrete containing coarse aggregate, but little or no fine aggregate [85]. This produces a material with a volume of large pores, which allow the movement of liquid water and oxygen downwards, and water vapour upwards. For permeable pavements to be successful, a soil which permits percolation of water is essential, since an impermeable soil will still lead to high moisture levels close to the surface [86].

The success of permeable pavements, however, is reported as being somewhat inconsistent, with levels of moisture and oxygen beneath test pavements showing little correlation between permeable and impermeable surfaces [87].

When compacted subgrade is located below pavements, there exists the possibility of including 'root paths' [86]. These are trenches in the sub-grade lined with strip drain material leading to the other side of the paved area. The trenches are backfilled with good quality soil before the rest of the pavement construction is completed. The theory behind this approach is that it offers a low-resistance path for the root to grow beneath the pavement into an area where growth can continue without disruption of infrastructure. Such systems are also sometimes referred to as 'root breakout zones' [58]. There is currently an absence of data with regards to the effectiveness of this approach.

The compaction of soils beneath pavements can be avoided through the use of concrete pavement slabs suspended over the soil, and thus not requiring any support from the underlying ground. The approach has been demonstrated as promoting growth of trees, but its effectiveness at preventing pavement damage is currently unclear [87].

The growth of roots away from trees is stimulated by the absence of either water or nutrients. If adequate quantities of both are provided, growth will be suppressed. This approach clearly needs a degree of planning and resources, and is probably best suited to urban areas. However, the introduction of tree pits has made this a more realistic option. Tree pits are concrete lined chambers located beneath the ground and covered by a suspended concrete slab [58]. The bottom of the pit consists of a bed of gravel directing water into a drain. The chamber holds the root ball of the tree, plus a volume of soil. The tree trunk projects through an opening in the concrete cover, and is often encircled by a cast iron grille. A pipe also runs through the interior of the chamber to deliver water and, possibly, nutrients. This pipe will deliver water to a series of tree pits, thus making irrigation relatively easy and automatable. The tree pit has the advantage not only of allowing irrigation and fertilization, but also allows for the introduction of root paths, avoids compaction through the suspended slab, and also permits control over the size of the chamber, which—as will be discussed later in this chapter—can be useful in limiting damage to pipes.

Tree roots and foundations

There are two effective approaches to limiting damage to foundations as a result of desiccation of soil by tree roots. Firstly, statistical analysis of instances of damage allows likely safe distances that buildings can be located away from specific tree species. For instance, the maximum distance from oak trees where 90% of the instances of damage were reported in the UK is 18 m [59]. This approach can also be applied when planting new trees. Secondly, where relocation of a planned structure is not possible, deeper foundations can be used to take the foundation below the zone of soil affected by the drying effects of the roots [60].

Pipes

The type of seal used at pipe joints will define the ease with which root intrusion can occur. The previous surveys of root intrusion into pipes have, in part, examined pipes with more traditional seals, such as textile strips impregnated with bitumen and then sealed in place with cement. This sealing technique has been superseded, firstly with natural rubber rings and secondly with synthetic rubber gaskets designed to making fitting easier. A study examining the more modern seal found that despite its composition and design, roots had little problem in breaching it [63]. Improvement was observed, however, when the outside of the joints was further sealed with self-vulcanising tape.

The recommendations which have arisen from this study are that care should be taken in planting trees near pipelines, with consideration of distances and tree species. Tree species should be those having a low 'root energy'—i.e., being less intrusive in nature. Common lime (*Tilia europaea*) is suggested as an example by the researchers. It was also proposed that the plant bed (the volume of soil which is excavated and replaced when a tree is planted) should be made larger than normal. The reason why such an approach is likely to be effective is that this volume of soil tends to be more porous. Root growth will typically initially be limited to this zone prior to infiltrating the denser surrounding soil, thus extending the period of time before roots reach the pipe. Finally, the use of root barriers geotextiles was proposed as a means of slowing the progress of roots, although it was stressed that such barriers will typically hinder root growth rather than stop it.

Where root intrusion has occurred, specialised cutting equipment can be deployed in pipes to remove blockages. Root intrusions can also be removed through the application of formulations containing the contact herbicide metam sodium and a growth inhibitor [86]. By using a contact herbicide, the intruding roots can be destroyed without killing the entire plant. However, since it requires releasing quantities of herbicide into wastewater, in many countries this approach is not permitted.

5.7 References

[1]   Lee RE (2008) Phycology, 4th Ed., Cambridge University Press, Cambridge.
[2]   Barberousse H, Brayner R, Botelho Do Rego AM, Castaing J-C, Beurdeley-Saudou P and Colombet J-F (2007) Adhesion of façade coating colonisers, as mediated by physico-chemical properties. Biofouling 23(1): 15–24.
[3]   Barberousse H, Ruiz G, Gloaguen V, Lombardo RJ, Djediat C, Mascarell G and Castaing J-C (2006) Capsular polysaccharides secreted by building façade colonisers: characterisation and adsorption to surfaces. Biofouling 22(6): 361–370.
[4]   Lisci M, Monte M and Pacini E (2003) Lichens and higher plants on stone: a review. International Biodeterioration and Biodegradation 51(1): 1–17.
[5]   English Heritage (2014) Landscape Advice Note: Vegetation on Walls. English Heritage, London.
[6]   Segal S (1969) Ecological Notes on Wall Vegetation. Junk, The Hague.
[7]   Kumar R and Kumar AV (1999) Biodeterioration of Stone in Tropical Environments—An Overview. The Getty Conservation Institute, Los Angeles, CA, USA.
[8]   Kew Royal Botanic Gardens (2008) Seed Information Database version 7.1 Kew Royal Botanic Gardens, London, UK. http://data.kew.org/sid/
[9]   Çalişkan O, Mavi K and Polat A (2012) Influences of presowing treatments on the germination and emergence of fig seeds (*Ficus carica* L.). Acta Scientiarum Agronomy 34(3): 293–297.
[10] Janick J and Paull RE (2008) The Encyclopedia of Fruit and Nuts. Centre for Agriculture and Biosciences International, Wallingford, UK.

[11] Hintze C, Heydel F, Hoppe C, Cunze S, König A and Tackenberg O (2013) D³: The dispersal and diaspore database—baseline data and statistics on seed dispersal. Perspectives in Plant Ecology, Evolution and Systems **15(3)**: 180–192. www.seeddispersal.info accessed November 2016.

[12] Algoplus (2016) Plant pH Preference List, Algoplus, Jacksonville, FL, USA. www.algoplus.net/soilpHinfo.html accessed November 2016.

[13] Ajuntament de Barcelona (2016) Trees and Palm Trees of Barcelona, Ajuntament de Barcelona, Barcelona, Spain. w110.bcn.cat/portal/site/MediAmbient/menuitem.37 ea1e76b6660e13e9c5e9c5a2ef8a0c/index1536.html?vgnextoid=5834c50d8a35b210Vgn VCM10000074fea8c0RCRD&vgnextchannel=5834c50d8a35b210VgnVCM10000074fea 8c0RCRD&vgnextrefresh=1&lang=en_GB accessed November 2016.

[14] Forest Ecology and Forest Management Group (2016) Tree factsheet—*Robinia pseudoacacia* L. Forest Ecology and Forest Management Group, Wageningen University, Wageningen, Netherlands.

[15] Centre for Agriculture and Biosciences International (2016) Invasive Species Compendium. www.cabi.org/isc/datasheet/14280 accessed November 2016.

[16] Trewartha J and Trewartha S (2005) Producing Capers in Australia—Viability Study. Australian Government Rural Industries Research and Development Corporation, Canberra, Australia.

[17] Wacquant JP and Picard JB (1992) Nutritional differentiation among populations of the Mediterranean shrub *Dittrichia viscosa* (Asteraceae) in siliceous and calcareous habitats. Oecologia **92(1)**: 14–22.

[18] Beerling DJ (1991) The testing of cellular concrete revetment blocks resistant to growths of *Reynoutria japonica* houtt (Japanese knotweed). Water Research **25(4)**: 495-498.

[19] Weisse T and Stadler P (2006) Effect of pH on growth, cell volume, and production of freshwater ciliates, and implications for their distribution. Limnology and Oceanography **51(4)**: 1708–1715.

[20] Bell TAS, Prithiviraj B, Wahlen BD, Fields MW and Peyton BM (2015) A lipid-accumulating alga maintains growth in outdoor, alkaliphilic raceway pond with mixed microbial communities. Frontiers in Microbiology **6**: 1480–1491.

[21] Tran TH, Govi A, Guyonnet R, Grosseau P, Lors C, Garcia-Diaz E, Damidot D, Devèse O and Ruote B (2012) Influence of the intrinsic characteristics of mortars on biofouling by *Klebsormidium flaccidum*. International Biodeterioration and Biodegradation **70**: 31–39.

[22] Štechová T, Hájek M, Hájková P and Navrátilová J (2008) Comparison of habitat requirements of the mosses *Hamatocaulis vernicosus*, *Scorpidium cossonii* and *Warnstorfia exannulata* in different parts of temperate Europe. Preslia **80(4)**: 399–410.

[23] Ogbimi AZ, Owoeye YB, Ibeyemi VO and Bofede AV (2014) Effects of pH, photoperiod and nutrient on germination and growth of *Calymperes erosum* C. Muell. Gemmaling. Journal of Botany **2014**: 1–5.

[24] Gillman JH, Dirr MA and Braman SK (1998) Effects of dolomitic lime on growth and nutrient uptake of *Buddleia davidii* 'Royal Red' grown in pine bark. Journal of Environmental Horticulture **16(2)**: 111–113.

[25] Hewitt EJ and Smith TA (1975) Plant Mineral Nutrition. English Universities Press, London, UK.

[26] Barberousse H, Tell G, Yéprémian C and Couté A (2006) Diversity of algae and cyanobacteria growing on building façades in France. Algological Studies **120**: 81–105.

[27] White PJ and Broadley MR (2003) Calcium in plants. Annals of Botany **92(4)**: 487–511.

[28] Rout G, Samantaray S and Das P (2001) Aluminium toxicity in plants: a review. Agronomie **21(1)**: 3–21.

[29] Taylor GJ (1991) Current views of the aluminum stress response: the physiological basis of tolerance. Current Topics in Plant Biochemistry and Physiology **10**: 57–93.

[30] Ryan PR, Delhaize E and Jones DL (2001) Function and mechanism of organic anion exudation from plant roots. Annual Review of Plant Physiology and Plant Molecular Biology **52**: 527–560.

[31] Ström L (1997) Root exudation of organic acids: importance to nutrient availability and the calcifuge and calcicole behaviour of plants. Oikos **80(3):** 459–466.

[32] Kobayashi T, Inamori Y, Tanabe I and Obayashi A (1974) Organic acid production by unicellular green algae during the heterotrophic growth. Nippon Nōgeikagaku Kaishi **48(4):** 261–267.

[33] Friedmann EI (1980) Endolithic microbial life in hot and cold deserts. Origins of Life and Evolution of Biospheres **10(3):** 223–235.

[34] Viles H (1995) Ecological perspectives on rock surface weathering: Towards a conceptual model. Geomorphology **13(1):** 21–35.

[35] Garcia-Rowe J and Saiz-Jimenez C (1991) Lichens and bryophytes as agents of deterioration of building materials in Spanish cathedrals. International Biodeterioration **28(1-4):** 151–163.

[36] Klavina L, Springe G, Nikolajeva V, Martsinkevich I, Nakurte I, Dzabijeva D and Steinberga I (2015) Chemical composition analysis, antimicrobial activity and cytotoxicity screening of moss extracts (moss phytochemistry). Molecules **20(9):** 17221–17243.

[37] Bates JW (1992) Mineral nutrient acquisition and retention by bryophytes. Journal of Bryology **17(2):** 223–240.

[38] Jayakumar S and Saravanane R (2010) Biodeterioration of coastal concrete structures by marine green algae. International Journal of Civil Engineering **8(4):** 352–361.

[39] Hughes P, Fairhurst D, Sherrington I, Renevier N, Morton LHG, Robery PC and Cunningham L (2013) Microscopic study into biodeterioration of marine concrete. International Biodeterioration and Biodegradation **79:** 14–19.

[40] Dubosc A, Escadeillas G and Blanc PJ (2001) Characterization of biological stains on external concrete walls and influence of concrete as underlying material. Cement and Concrete Research **31(11):** 1613–1617.

[41] De Muynck W, Ramirez AM, De Belie N and Verstaete W (2009) Evaluation of strategies to prevent algal fouling on white architectural and cellular concrete. International Biodeterioration and Biodegradation **63(6):** 679–689.

[42] Barberousse H, Ruot B, Yéprémian C and Boulon G (2007) An assessment of façade coatings against colonisation by aerial algae and cyanobacteria. Building and Environment **42(7):** 2555–2561.

[43] Fonseca AJ, Pina F, Macedo MF, Leal N, Romanowska-Deskins A, Laiz L, Gómez-Bolea A and Saiz-Jimenez C (2010) Anatase as an alternative application for preventing biodeterioration of mortars: Evaluation and comparison with other biocides. International Biodeterioration and Biodegradation **64(5):** 388–396.

[44] Crispim CA, Gaylarde PM and Gaylarde CC (2003) Algal and cyanobacterial biofilms on calcareous historic buildings. Current Microbiology **46(2):** 79–82.

[45] Rosato VG (2008) Pathologies and biological growths on concrete dams in tropical and arid environments in Argentina. Materials and Structures **41(7):** 1327–1331.

[46] Andersson MH, Berggren M, Wilhelmsson D and Öhman MC (2009) Epibenthic colonization of concrete and steel pilings in a cold-temperate embayment: a Weld experiment. Helgoland Marine Research **63(3):** 249–260.

[47] Olivia M, Moheimani N, Javaherdashti R, Nikraz HR and Borowitzka MA (2013) The influence of micro algae on corrosion of steel in fly ash geopolymer concrete: a preliminary study. Advanced Materials Research **626:** 861–866.

[48] Rindi F, Guiry MD, Critchley AT and Ar Gall E (2003) The distribution of some species of Trentepohliaceae (Trentepohliales, Chlorophyta) in France. Cryptogamie Algologie **24(2):** 133–144.

[49] Jukonienė I (2008) The impact of anthropogenic habitats on rare bryophyte species in Lithuania. Folia Cryptogamica Estonica **44:** 55–62.

[50] Ignatov MS, Ignatova EA, Akatova TV and Konstantinova NA (2002) Bryophytes of the Khosta' Taxus and Buxus forest (Western Caucasus, Russia). Arctoa **11(1):** 205–214.

[51] Sakovich A and Rykovsky G (2011) Comparative analysis of the bryophyte floras of Northwest Belarus concrete fortification and Carpathians. Biodiversity Research and Conservation **24(1):** 23–27.

[52] Gregory PJ (1988) Growth and functioning of plant roots. *In*: Wild A (ed.). Russell's Soil Conditions and Plant Growth. Longmans, London, UK.

[53] Raphael JM (1984) Tensile strength of concrete. Journal of the American Concrete Institute **81(2):** 158–165.

[54] Viles H, Sternberg T and Athersides A (2011) Is ivy good or bad for historic walls? Journal of Architectural Conservation **17(2):** 25–41.

[55] Elinç ZK, Korkut T and Kaya LG (2013) *Hedera helix* L. and damages in Tlos ancient city. International Journal of Development and Sustainability **2(1):** 333–346.

[56] Sternberg T, Viles H, Cathersides A and Edwards M (2010) Dust particulate absorption by ivy (*Hedera helix* L.) on historic walls in urban environments. Science of the Total Environment **409(1):** 162–168.

[57] Lewin SZ and Charola AE (1981) Plant life on stone surfaces and its relation to stone conservation. Scanning Electron Microscopy **1981:** 563–568.

[58] Roberts J, Jackson N and Smith M (2013) Tree Roots in the Built Environment. Arboricultural Association, Stonehouse, UK.

[59] Building Research Establishment (1999) Digest 298: Low-Rise Building foundations: the Influence of Trees in Clay Soils. Building Research Establishment, Watford, UK.

[60] National House Building Council (2016) NHBC Standards 2017. National House Building Council, Milton Keynes, UK.

[61] Kuliczkowska E and Parka A (2017) Management of risk of environmental failure caused by tree and shrub root intrusion into sewers. Urban Forestry and Urban Greening **21:** 1–10.

[62] Östberg J, Martinsson M, Stål Ö and Fransson A-M (2012) Risk of root intrusion by tree and shrub species into sewer pipes in Swedish urban areas. Urban Forestry and Urban Greening **11(1):** 65–71.

[63] Ridgers D, Rolf K and Stål Ö (2006) Management and planning solutions to lack of resistance to root penetration by modern PVC and concrete sewer pipes. Arboricultural Journal **29(4):** 269–290.

[64] Pohls O, Bailey NG and May PB (2004) Study of root invasion of sewer pipes and potential ameliorative techniques. Acta Horticulturae **643:** 113–121.

[65] Garcia-Rowe J and Saiz-Jimenez C (1991) Lichens and bryophytes as agents of deterioration of building materials in Spanish Cathedrals. International Biodeterioration **28(1-4):** 151–163.

[66] Javaherdashti R, Nikraz H, Borowitzka M, Moheimani N and Olivia M (2009) On the impact of algae on accelerating the biodeterioration/biocorrosion of reinforced concrete: a mechanistic review. European Journal of Scientific Research **36(3):** 394–406.

[67] Jayakumar S and Saravanane R (2010) Detrimental effects on coastal concrete by *Ulva fasciata*. Construction Materials **163(4):** 239–246.

[68] Coombes MA, Naylor LA, Viles HA and Thompson RC (2013) Bioprotection and Disturbance: seaweed, microclimatic stability and conditions for mechanical weathering in the intertidal zone. Geomorphology **202:** 4–14.

[69] Hughes P, Fairhurst D, Sherrington I, Renevier N, Morton LHG, Robery PC and Cunningham L (2013) Microscopic examination of a new mechanism for accelerated degradation of synthetic fibre reinforced marine concrete. Construction and Building Materials **41:** 498–504.

[70] Hughes P, Fairhurst D, Sherrington I, Renevier N, Morton LHG, Robery PC and Cunningham L (2014) Microbial degradation of synthetic fibre-reinforced marine concrete. **86(A):** 2–5.

[71] Suseela M and Toppo K (2007) Algal biofilms on polythene and its possible degradation. Current Science **92(3):** 285–287.

[72] Lane RA (2005) Under the microscope: understanding, detecting, and preventing microbiologically influenced corrosion. Journal of Failure Analysis and Prevention **5(10-12):** 33–38.

[73] Qin J (2005) Bio-Hydrocarbons from Algae: Impacts of temperature, light and salinity on algae growth. Australian Government Rural Industries Research and Development Corporation, Barton, Australia.

[74] Barberousse H, Lombardo RJ, Tell G and Couté A (2006) Factors involved in the colonisation of building façades by algae and cyanobacteria in France. Biofouling **22(2):** 69–77.

[75] Singh SP and Singh P (2015) Effect of temperature and light on the growth of algae species: A review. Renewable and Sustainable Energy Reviews **50:** 431–444.

[76] Navío JA, Macias M, Garcia-Gómez M and Pradera MA (2008) Functionalisation versus mineralisation of some N-heterocyclic compounds upon UV-illumination in the presence of un-doped and iron-doped TiO_2 photocatalysts. Applied Catalysis B: Environmental **82(3-4):** 225–232.

[77] Maury-Ramirez A, De Muyncka W, Stevens R, Demeestere K and De Belie N (2013) Titanium dioxide based strategies to prevent algal fouling on cementitious materials. Cement and Concrete Composites **36:** 93–100.

[78] Linkous CA, Carter GJ, Locuson DB, Ouellette AJ, Slattery DK and Smith LA (2000) Photocatalytic inhibition of algae growth using TiO_2, WO_3, and co-catalyst modifications. Environmental Science and Technology **34(22):** 4754–4758.

[79] Building Research Establishment (1992) BRE Digest 370: Control of Lichens, Moulds and Similar Growths. Building Research Establishment, Watford, UK.

[80] The Concrete Society (2013) Visual Concrete—Weathering, Stains and Efflorescence. The Concrete Society, Camberley, UK.

[81] De Graef B, Dick J, De Windt W, De Belie N and Verstraete W (2004) Cleaning of concrete fouled by algae with the aid of thiobacilli. pp. 55–64. *In*: Silva MR (ed.). Second International RILEM Workshop on Microbial Impact on Building Materials. RILEM, Paris, France.

[82] Dyer TD (2014) Concrete Durability. CRC Press, Boca Raton, FL, USA.

[83] Building Research Establishment (1996) BRE Digest 418: Bird, Bee and Plant Damage to Buildings. Building Research Establishment, Watford, UK.

[84] Smiley ET, Key A and Greco C (2000) Root barriers and windthrow potential. Journal of Arboriculture **26(4):** 213–217.

[85] Tennis PD, Leming ML and Akers DJ (2004) Pervious Concrete Pavements. Portland Cement Association, Skokie, IL, and National Ready Mixed Concrete Association, Silver Spring, MD, USA.

[86] Watson GW, Hewitt AM, Custic M and Lo M (2014) The management of tree root systems in urban and suburban settings II: a review of strategies to mitigate human impacts. Arboriculture and Urban Forestry **40(5):** 249–271.

[87] Smiley ET, Calfee L, Fraedrich BR and Smiley EJ (2006) Comparison of structural and noncompacted soils for trees surrounded by pavement. Arboriculture and Urban Forestry **32(4):** 164–169.

Chapter 6

Damage to Concrete from Animal Activity

6.1 Introduction

This chapter examines the impact animal activity has on the durability of concrete. Animals are members of the kingdom Animalia. They are eukaryotes, like fungi, but differ significantly in that they are exclusively multicellular and are motile. This chapter differs from previous chapters in that, given the variety of organisms within this kingdom, the complexity of their life-cycles, plus the relatively small size of the sub-set which cause deterioration of concrete, there is little benefit in understanding the organisms themselves in detail. Thus, the approach taken will be to identify types of animal documented as having caused damage to concrete structures, with a discussion of approaches which may protect structures vulnerable to such attack.

Additionally, the chapter focuses largely on biodeterioration resulting from concrete being specifically targeted by an animal to satisfy some requirement for survival. For instance, whilst it is likely that abrasion of concrete surfaces by the hooves of cattle compromises the durability of concrete worldwide, the cattle do not benefit from this abrasion and the concrete is not specifically selected by the animals because it can be abraded, or contains something that they need to live.

The majority of this chapter involves the marine animals, since these are the most damaging to concrete, albeit under specific conditions. In most cases the manner in which these organisms attack concrete is comparable, and common solutions exist.

The chapter ends with a discussion of the potential for damage to concrete from bird droppings. Whilst this falls outside of the chapter remit defined above, this topic has been the subject of some speculation in recent years, and this discussion aims to evaluate the facts and draw balanced conclusions from them.

6.2 Marine Animals

Submerged concrete surfaces in a marine environment will typically rapidly acquire an attachment of marine life. An example of how mixed communities of marine organisms develop on a concrete pile located off the West coast of Sweden is shown in Figure 6.1. Whilst the manner in which colonization occurs is dependent on geography, the sequence is reasonably typical. Initially the surface is colonized by algae and the genus *Laomedea* —small plant-like animals which are members of the Hydrozoa class of animals. The surface gradually becomes more diverse with the appearance of *Ciona intestinalis* (a sponge) and at least two species of ascidians, or sea squirts: a species of the genus *Ascidiella* and *Corella parallelogramma*. Lesser coverage by the calcareous tube worm *Pomatoceros triqueter* was briefly observed, but this species is seemingly gone after 12 months. These reside against the surface within a calcium carbonate tube which the worm forms around itself. Finally, the appearance of barnacles in the form of *Balanus improvisus* was observed.

This assemblage of marine species is far from comprehensive. Other organisms which may be found growing on submerged marine concrete can include the Bryozoa ('moss animals'), hard and soft corals, anemones, sea urchins and various molluscs [2, 3]. Molluscs will typically appear more than 12 months after placing.

From this range of organisms, only a few are known to cause deterioration of concrete. Specifically, various molluscs and sponges have

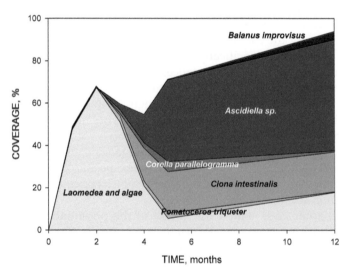

Figure 6.1 Colonisation of a concrete pile off the West coast of Sweden [1].

been documented as causing damage. It has also been found that some marine worms and sea urchins may also be capable of causing damage.

It has been observed previously that, in some cases, the colonization of a concrete surface may at least partly be determined by the presence of substances in the material which can be used as nutrients. In the case of marine organisms, this is not the reason for causing damage—all of the nutrients required by such marine animals are likely to be present in the seawater itself either as dissolved chemical species or as other organisms or particles of debris. Instead, in most cases, these animals are boring beneath the surface for shelter, or the damage is a side-effect of feeding off other organisms on the concrete surface.

Before examining each of these types of marine animal in more detail, it is worth briefly mentioning the barnacle, whose presence on concrete would appear to have potentially beneficial effects with regards to durability.

Barnacles are arthropods which fix themselves permanently to a hard substrate, feeding on particles suspended in seawater. They are encased in a carapace which is composed of calcium carbonate. Most barnacle species are attached to the substrate by a calcium carbonate basal plate or membrane, which effectively isolates the animal from the substrate.

The basal plate is typically a relatively dense formation, and where barnacles are present on concrete, it has the tendency to make the surface less permeable to ingressing chemical species. This has particular benefits from a concrete durability perspective in limiting the movement of chloride ions from seawater through the concrete cover towards steel reinforcement, where their presence will eventually initiate corrosion. The effect of barnacle coverage on the chloride diffusion coefficient of concrete is shown in Figure 6.2.

Barnacle colonization of concrete tends to be slower than for other surfaces, such as steel [1]. This is largely the result of the high initial pH [5].

6.2.1 Molluscs

A number of members of the phylum Mollusca are capable of either boring into or abrading the surface of concrete. The most frequently documented instances of mollusc damage to concrete involve the bivalves.

Bivalves

Certain bivalves bore a cylindrical borehole into the substrate (usually rock) and live in the resulting hole, enlarging it as the organism grows.

Table 6.1 lists bivalve species which have been observed to bore into concrete. One location in this table is the Panama Canal, where *Lithophaga aristata* was established as causing damage to concrete piers. However, the same report also suggests that other species in this location may also be

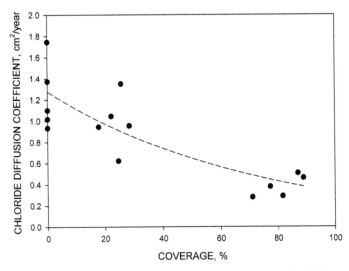

Figure 6.2 Chloride diffusion coefficients of concrete covered to different extents by barnacle communities [4].

Table 6.1 Bivalve species confirmed as boring into concrete.

Genus	Species	Order/Family	Boring Mechanism	Location	References
Lithophaga	–	**Mytiloida/** Mytilidae	Chemical	Saudi Arabia/ Jamaica	[6, 7]
	aristata			Panama Canal	[8]
Pholadidea	*penita*	**Myoida/** Pholadidae	Mechanical	Los Angeles Harbour	[8]
Platyodon	*cancellatus*	**Myoida/**Myidae	Mechanical	Los Angeles Harbour	[8]

responsible, namely *Carditamera affinis* and *Hiatella solida* [8]. Additionally, a review paper on marine borers also identifies the *Lithophaga, Pholadidea, Platyodon* and *Hiatella* genera as being known to bore into concrete, and adds *Zirfaea, Petricola,* and *Carditamera* to this [9].

Most bivalves use mechanical action to bore into rock or concrete. This is achieved using a filing/rasping action of the shell to remove material. *Lithophaga*, however, does not have a shell suitable for mechanical boring and secretes a substance from a gland known as the pallial gland which contains a mucoprotein capable of chelating calcium (see Chapter 2) [10]. As a result, *Lithophaga* are only capable of boring into calcareous rock such as limestone. The high calcium content of the cement matrix of concrete makes it a suitable substrate for *Lithophaga* boring.

In many instances, observation of mechanical borers in concrete have been associated with poor mechanical properties in the material. For instance, in the case of *Pholadidea* and *Platyodon* (Figure 6.3), the concrete was found to be extremely weak, which was blamed on the manner in which the concrete had been placed underwater leading to a high water/cement ratio [8]. However, it would appear that the strength of concrete is not a barrier when the boring mechanism is chemical: in a study of concrete piles off the coast of Saudi Arabia *Lithophaga* were found to be able to penetrate concrete piles with a strength of 50.7 MPa [6].

In this case, the diameter of the boreholes was found to be around 5 mm and the holes extended up to 15 mm into the concrete. The species can bore at a rate of up to 10 mm per year and the diameter of boreholes can be as large as 15 mm for a mature animal [11].

It is notable that, where stated, concrete attacked by *Lithophaga* contained calcareous aggregate [6, 7]. Thus, the entire mass of concrete was presumably wholly vulnerable to the type of chemical boring employed by *Lithophaga*. Indeed *Lithophaga* has been observed to bore through cement and calcareous aggregate with equal ease [7]. The same study, involving a concrete block 'reef' colonised by *Lithophaga* (albeit alongside other boring organisms) had encountered annual rates of bioerosion of around 4.5% by

Figure 6.3 Bivalves of the genus Pholadidea in concrete taken from concrete pile jackets off the Californian coast [8].

volume. Based on the descriptions of damage made by the researchers, the majority was from the bivalve.

The number of reported cases of this type of biodeterioration is very small, and it is likely that concrete containing siliceous aggregate—which cannot be removed by the chemical boring mechanism—is more resistant to *Lithophaga*.

Despite their name, shipworms are bivalve molluscs. They belong to the genera *Tenera* and *Banksia* which both belong to the family Teredinidae and the order Teleodesmacea. Their name refers to their ability to bore into timber (historically, that of ships' hulls) to make their home, and their elongated body with two small shells located at their head. These shells are used as rasping tools for boring. One literature source states that shipworms have been known to bore into concrete, although no specific evidence is provided [9]. Given that these animals utilize a mechanical boring mechanism, it is reasonable to suppose that —as for the other bivalves that bore in this way—only weaker and/or deteriorated concrete is likely to be vulnerable.

Chitons and other grazing molluscs

Chitons are molluscs which live on rock surfaces in marine environments, often in intertidal or subtidal zones. They are covered in eight articulated shell plates which protect them, whilst giving them considerable freedom of movement. Chitons are able to move over rock surfaces, feeding on algae and other organisms, such as bryozoans. They are equipped with a rasp known as a radula which is used to scrape these organisms from the surface. The radula is coated in a layer of extremely hard magnetite (Fe_3O_4) and the strategy the chiton employs to maximize the efficiency of grazing is to abrade not only the food source, but also a thin layer of the underlying rock. The particles of rock pass through the animal's digestive system.

Chitons are commonly found on concrete surfaces in coastal locations. Examples include the species *Sypharochiton pelliserpentis* on seawalls in Sydney Harbour [12] and *Mopalia hindsii* on concrete piles in Monterey Harbor in the United States [13]. Whilst there appear to be no estimates of abrasion rates for chitons feeding from concrete surfaces, abrasion rates for limestone have been estimated. Two studies of chitons (*Acanthopleura gemmata*) living on limestone around the coast of an island in the Great Barrier Reef estimated erosion rates of 0.2–0.7 mm/year and 0.16 mm/year [14, 15].

These rates were considered significant by the researchers whose primary interest was the effect of bioerosion processes on coral reefs. However, assuming that concrete would typically display erosion rates at the lower ends of this scale, due to the presence of more resistant aggregate particles, rates are likely to be considerably lower than those observed for bivalves such as *Lithophaga*. Thus, it is unlikely that chitons present a significant threat to concrete durability.

Other grazing molluscs include the gastropods such as limpets (e.g., members of the families Nacellidae, Lottiidae, Siphonariidae and Fissurellidae), sea snails (e.g., Trochidae, Littorinidae and Neritidae). Limpets possess radula with silica-rich teeth, which are also implicated in bioerosion processes on rocks [16]. However, rates of erosion would not appear to exceed those of chitons. Rates from the grazing of sea snails are thought to be even lower.

6.2.2 Sponges

Like the mollusc *Lithophaga*, boring sponges (phylum Porifera, class Demospongiae) utilise chemical secretions to dissolve calcium in rock substrates. Where this has occurred in concrete the effect has been the formation of networks of holes around 1 mm in diameter (Figure 6.4). These honeycomb-like structures are used by the sponge to anchor itself to rock surfaces, and can increase in depth at a rate of around 1 mm per year [17].

Table 6.2 lists reported instances of concrete attack by boring sponges. In all cases, the coarse aggregate used was calcareous, and in most instances it is noted that boring was entirely limited to the aggregate [6, 7, 17].

Whilst the borehole diameter and rate of increase in depth resulting from boring sponges is notably less than the values for *Lithophaga*, rates of

Figure 6.4 Holes formed in calcareous aggregate in concrete by a boring sponge [6].

Table 6.2 Boring sponges found in concrete structures.

Genus	Species	Order/Family	Location	Reference
Cliona	–	Hadromerida/Clionaidae	Saudi Arabia	[6]
	–		Jamaica	[7]
	–		Off coast of Georgia, USA	[17]
Damiria	–	Poecilosclerida/Acarnidae	Jamaica	[7]

deterioration have the potential to be significant, if coverage of a surface is complete. One study examining the erosion rate of coral reefs which had been colonised by boring sponges estimated rates between 0.84 and 23.0 kg $CaCO_3/m^2$year [18] (where the area term in the units is the surface area of the coral) with the highest rate being observed for colonies of the species *Cliona lampa*. However, it is unlikely that coral is wholly comparable to calcareous aggregate in concrete, since it has a very different microstructure, and the presence of cement around the aggregate may hamper boring rates. Thus, the lower end of this range is likely to be more representative of concrete.

Boring sponges tend to be at their most destructive in regions close to the equator. It has been found that whilst temperature does not have a direct influence on the rate of boring [19], the productivity of the water in which the sponge lives does [7], presumably because there is a readily available supply of nutrients. Interestingly, a study looking at the boring of sponges into the shells of molluscs has found that, as the pH declines, the rate of boring increases [19]. This has potentially significant implications, since reduced pH of the oceans is predicted as a result of climate change.

6.2.3 Sea urchins

Some sea urchins are capable of making quite substantial burrows in rock. The burrow serves to provide shelter, and is formed by the animal securing itself to the rock with its feet and abrading the surface with its five jaws and rubbing with its test (shell) and spines. The motion is usually back and forth in only one direction, with the ultimate effect of producing a burrow which consists of a trench which is triangular in profile, with the urchin normally located at the lowest point [16].

The purple sea urchin (*Strongylocentrotus purpuratus*)—which can have a diameter of around 65 mm—has been reported to have burrowed to significant depths into concrete breakwaters on the Californian coast [20].

6.2.4 Marine worms

Another marine borer identified as being capable of boring into concrete are the marine worms of the class Polychaeta. Many of these worms are capable of boring into calcareous rock. Initially it was believed that this was achieved through a combination of chemical and mechanical processes achieved through the use of calcium-chelating mucopolysaccharide secretions from glands and the use of brush-like structures known as setae [21]. However, experiments in which the setae were removed found that the worm could bore without difficulty into calcareous rock, presumably through purely chemical means [22].

A study which examined the biodeterioration of concrete blocks in an artificial reef-type structure found that the worms appeared to favour boring into the cement matrix rather than into the calcareous aggregate used in the blocks [7].

6.2.5 Limiting damage from marine animals

It is firstly worth noting that the most aggressive forms of marine animal attack are limited to warmer zones, presumably where the productivity of the sea is sufficiently high to support the sort of energy requirements of boring on the scales reported. Thus, damage from marine animals in many parts of the world is likely to be on a scale which is of little concern.

The mechanisms employed by marine animals to bore or burrow into concrete fall into two categories: mechanical and chemical, and different options are available in limiting damage depending on what mechanism is used.

The results discussed in this section point to the likelihood that using non-calcareous aggregate in concrete is likely to limit the magnitude of attack. However, different animal types showed different preferences. Thus, boring sponges, which seem to exclusively bore into calcareous aggregate would probably be severely limited in its ability to bore by the presence of wholly siliceous aggregate. *Lithophaga* appears to be able to bore through both cement and calcareous aggregate. Nonetheless, use of non-calcareous aggregate might be expected to hamper or entirely prevent boring. This appears to be supported by the apparent complete absence of reported incidences of *Lithophaga* in concrete without calcareous aggregate.

The only detailed report of Polychaeta worms boring into concrete appears to indicate less sensitivity to aggregate type and a possible preference for the cement matrix, indicating that sourcing specific aggregate types to control this form of attack may not be an option.

Where mechanical damage is observed, it is often associated with concrete of a low quality, and simply achieving appropriate performance through mix design and good practice in mixing, placing and curing is likely to limit biodeterioration considerably.

Whether the mechanism used by marine animals is chemical or mechanical, the use of physical barriers to prevent boring is likely to be an effective option. The report of attack by *Lithophaga* and *Cliona* off the coast of Saudi Arabia examined possible barrier options [6]. These included polymer surface wraps, underwater epoxy paints, polyester bags deployed around the concrete structure surface into which cement grout is pumped, and moulded fiberglass jackets. The solution that was eventually selected was the jackets, which were secured using an epoxy grout introduced into the annulus between the jacket and the concrete surface.

Barriers of these types are effective through the provision of an additional layer of protection between the concrete surface and the seawater. However, it should be noted that purple sea urchins have been reported to have abraded through steel casings around concrete piles [20], indicating that even a strong barrier may not entirely resolve the problem. How much of this damage was carried out by the sea urchin and how much was the result of abiotic corrosion—which would be likely to be occurring in parallel—is not clear.

6.3 Deterioration from Exposure to Bird Droppings

Contact with bird droppings is sometimes cited as a source of damage to concrete and other construction materials [23]. The reason usually cited is that they contain acid. Bird droppings contain uric acid, present as crystals dispersed in a saturated solution of the compound—its solubility is low: around 0.068 g/l [24], although it becomes more soluble at higher pH [25]. It can also exist in a dihydrate form [26], which is slightly less soluble at 25°C. The compound is a relatively weak diprotic acid, having a pK_{a1} value of 5.4 and a pK_{a2} of 11.3 [27]. Uric acid forms a calcium salt—(calcium hydrogen urate hexahydrate, $Ca(C_5H_3N_4O_3)_2.6H_2O$). This is of low solubility (log K_{sp} at 25°C = –9.81) [28], and so contact between uric acid and cement may lead to the precipitation of the salt. This salt has a high molar volume (267 cm^3/ mol) [29], indicating the possibility of damage to concrete.

There is a lack of data regarding complex formation by the urate and hydrogen urate ions, with the exception of iron, and so prediction of the likely outcome of contact between cement and bird droppings must be done without this data. However, using geochemical modelling techniques (Figure 6.5) it is evident that the calcium urate salt is in fact likely to be precipitated at a location just *outside* the cement paste itself. Thus, deterioration by the mechanism of expansive salt precipitation is unlikely. Damage through conventional acidolysis still occurs, but is limited significantly by the lower concentrations of acid—the pH of bird droppings is typically only slightly below 7 [30].

A study examining the effect of solutions leached from bird droppings on crushed concrete samples held in test tubes for a period of five weeks found slightly higher levels of mass loss in some cases compared to a control

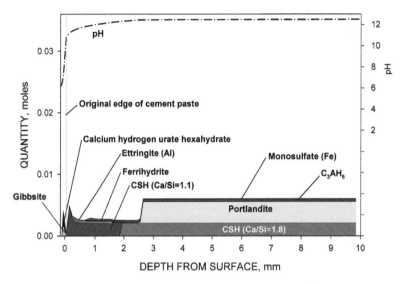

Figure 6.5 Quantities of solid phases versus depth obtained from geochemical modelling of uric acid attack of hydrated Portland cement. Model conditions: Excess of solid uric acid located in acid solution around cement paste; volume of acid solution = 4 l; mass of cement 80 g; diffusion coefficient = 5 × 10^{-13} m^2/s.

exposed to distilled water [30]. However, the differences were of the order of tenths of a percent, and the influence of such a difference on concrete performance is likely to be small. It should also be pointed out that by using leached solutions, the concentration of uric acid would have been low, due to its low solubility. Thus, there exists the possibility that the damaging effect of uric acid would have been greater if solid uric acid had been retained, since it would continue to dissolve during the experiments to replace any which had reacted with the concrete, as would happen in reality.

The chemical effect of concrete to bird droppings and uric acid specifically is still an area that needs further investigation, but there is little evidence of there being a serious problem.

However, one manner in which bird droppings may cause deterioration is as a source of nutrients for other species—such as fungi, bacteria and plants—which *can* damage concrete, in part through their production of other organic acids [31]. When fungi metabolize uric acid, the end products are urea and glyoxylic acid [32]. Urea does not affect the durability of concrete. Whilst experimental data relating to the interaction of glyoxylic acid with concrete is not available, it is likely that the compound behaves in a similar manner to other monocarboxylic acids, such as acetic acid (see Chapter 3).

It should be stressed that there are many other reasons for limiting the deposition of bird droppings on the built environment, including those of health and aesthetics. These reasons should, therefore, take priority in deciding whether measures are required. There exists much guidance on this issue, including a document by the Building Research Establishment in the UK [23].

6.4 References

[1] Andersson MH, Berggren M, Wilhelmsson D and Öhman MC (2009) Epibenthic colonization of concrete and steel pilings in a cold-temperate embayment: a field experiment. Helgoland Marine Research **63(3)**: 249–260.

[2] Oil and Gas UK (2013) The Management of Marine Growth During Commissioning. Oil and Gas UK, Aberdeen UK.

[3] Agatsuma Y (2013) Chapter 15: Stock enhancement. pp. 213–224. *In*: Lawrence JM (ed.). Sea Urchins: Biology and Ecology, 3rd Ed. Academic Press, London.

[4] Yokota H, Hamada H and Iwanami M (2013) Evaluation and prediction on performance degradation of marine concrete structures. *In*: Third International Conference on Sustainable Construction Materials and Technologies, Kyoto, Japan.

[5] Ido S and Shimrit P-F (2015) Blue is the new green—Ecological enhancement of concrete based coastal and marine infrastructure. Ecological Engineering **84**: 260–271.

[6] Wiltsie EA, Brown WC and Al-Shafei K (1984) Rock borer attack on Juaymah trestle concrete piles. pp. 767–772. *In*: First International Conference on Case Histories in Geotechnical Engineering, Missouri University of Science and Technology, Rolla, MO, USA.

[7] Scott PJB, Moser KA and Risk MJ (1988) Bioerosion of concrete and limestone by marine organisms: A 13 year experiment from Jamaica. Marine Pollution Bulletin **19(5)**: 219–222.

[8] Atwood WG and Johnson AA (1924) Marine Structures—their Deterioration and Preservation. National Research Council, Washington DC, USA.

[9] Castagna M (1973) Shipworms and other marine borers. Marine Fisheries Review **35(8)**: 7–12.

[10] Jaccarini V, Bannister WH and Micallef H (1968) The pallial glands and rock boring in Lithophaga lithophaga (Lamellibranchia, Mytilidae). Journal of Zoology **154(4)**: 397–401.

[11] Warme JE and Marshall NF (1969) Marine borers in calcareous terrigenous rocks of the pacific coast. American Zoologist **9(3)**: 765–774.

[12] Moreira J (2006) Patterns of occurrence of grazing molluscs on sandstone and concrete seawalls in Sydney Harbour (Australia). Molluscan Research **26(1)**: 51–60.

[13] Tucker JS and Giese AC (1959) Shell repair in chitons. Biological Bulletin **116(2)**: 318–322.

[14] Trudgill ST (1983) Preliminary estimates of intertidal limestone erosion, One Tree Island, southern Great Barrier Reef, Australia. Earth Surface Processes and Landforms **8(2)**: 189–193.

[15] Barbosa SS, Byrne M and Kelaher BP (2008) Bioerosion caused by foraging of the tropical chiton *Acanthopleura gemmata* at one tree reef, southern Great Barrier Reef. Coral Reefs **27(3)**: 635–639.

[16] Kazmer M and Taborosi D (2012) Bioerosion on the small scale—examples from the tropical and subtropical littoral. Hantkeniana **7(7)**: 37–94.

[17] Holland RB, Kurtis KE, Moser RD, Kahn LF, Aguayo F and Singh PM (2014) Multiple deterioration mechanisms in coastal concrete piles: A forensic case study. Concrete International **36(7)**: 45–51.

[18] Nava H and Carballo JL (2008) Chemical and mechanical bioerosion of boring sponges from Mexican Pacific coral reefs. The Journal of Experimental Biology **211(17)**: 2827–2831.

[19] Duckworth AR and Peterson BJ (2013) Effects of seawater temperature and pH on the boring rates of the sponge *Cliona celata* in scallop shells. Marine Biology **160(1)**: 27–35.

[20] Hinton S (1987) Seashore Life of Southern California: An Introduction to the Animal Life of California Beaches South of Santa Barbara. University of California Press, Oakland, CA.

[21] Dorsett DA (1961) The behaviour of Polydora ciliata (Johnst.). Tube-building and burrowing. Journal of the Marine Biological Association of the United Kingdom **41(3)**: 577–590.

[22] Haigler SA (1969) Boring mechanism of *Polydora websteri* inhabiting *Crassostrea virginica*. American Zoologist **9(3)**: 821–828.

[23] Building Research Establishment (1996) BRE Digest 418: Bird, Bee and Plant Damage to Buildings. Building Research Establishment, Watford, UK.

[24] Maiuolo J, Oppedisano F, Gratteri S, Muscoli C and Mollace V (2016) Regulation of uric acid metabolism and excretion. International Journal of Cardiology **213**: 8–14.

[25] Iwata H, Nishio S, Yokoyama M, Matsumoto A and Takeuchi M (1989) Solubility of uric acid and supersaturation of monosodium urate: why is uric acid so highly soluble in urine? Journal of Urology **142(4)**: 1095–8.

[26] Grases F, Villacampa AI, Costa-Bauzá A and Söhnel O (1999) Uric acid calculi. Scanning Microscopy **13(2-3)**: 223–234.

[27] Dawson RMC, Elliott DC, Elliott WH and Jones KM (1986) Data for Biochemical Research (3rd Ed.) Oxford University Press, Oxford, UK.

[28] Babić-Ivančić V, Füredi-Milhofer H and Brničević N (1992) Precipitation and solubility of calcium hydrogenurate hexahydrate. Journal of Research of the National Institute of Standards and Technology **97(3)**: 365–372.

[29] Presores JB, Cromer KE, Capacci-Daniel C and Swift JA (2013) Calcium urate hexahydrate. Crystal Growth and Design **13(12)**: 5162–5164.

[30] Huang CP and Lavenburg G (2011) Impacts of Bird Droppings and Deicing Salts on Highway Structures: Monitoring, Diagnosis, Prevention. Delaware Center for Transportation, University of Delaware, Newark, Delaware, USA.

[31] Bassi M and Chiantante D (1976) The role of pigeon excrement in stone biodeterioration. International Biodeterioration Bulletin **12(3)**: 73–79.

[32] Vera-Ponce de León A, Sanchez-Flores A, Rosenblueth M and Martínez-Romero E (2016) Fungal community associated with *Dactylopius* (Hemiptera: Coccoidea: Dactylopiidae) and its role in uric acid metabolism. Frontiers in Microbiology 9 Article 954.

Index

Printed and bound by CPI Group (UK) Ltd, Croydon, CR0 4YY

01/11/2024

01782624-0003